AF782187

Principles of Soil Mechanics

Principles of Soil Mechanics

Editor: Ian Menard

Callisto Reference
www.callistoreference.com

Callisto Reference,
118-35 Queens Blvd., Suite 400,
Forest Hills, NY 11375, USA

Visit us on the World Wide Web at:
www.callistoreference.com

© Callisto Reference, 2018

This book contains information obtained from authentic and highly regarded sources. All chapters are published with permission under the Creative Commons Attribution Share Alike License or equivalent. A wide variety of references are listed. Permissions and sources are indicated; for detailed attributions, please refer to the permissions page. Reasonable efforts have been made to publish reliable data and information, but the authors, editors and publisher cannot assume any responsibility for the vailidity of all materials or the consequences of their use.

ISBN: ISBN 978-1-64116-002-5 (Hardback)

The publisher's policy is to use permanent paper from mills that operate a sustainable forestry policy. Furthermore, the publisher ensures that the text paper and cover boards used have met acceptable environmental accreditation standards.

Trademark Notice: Registered trademark of products or corporate names are used only for explanation and identification without intent to infringe.

Printed in China.

Cataloging-in-Publication Data

Principles of soil mechanics / edited by Ian Menard.
p. cm.
Includes bibliographical references and index.
ISBN ISBN 978-1-64116-002-5
1. Soil mechanics. 2. Soil physics. I. Menard, Ian.
TA710 .P75 2018
624.151 36--dc23

Table of Contents

Preface

Soil mechanics studies the behavior and attributes of soil. The subject of soil mechanics helps in developing an understanding of buildings or structures that are supported by soil. The strength of the soil can be measured by the friction and aggregation found in it. It plays a crucial role in geotechnical engineering. For all those who are interested in soil mechanics, this book can prove to be an essential guide. Constant effort has been made to make the understanding of the concepts of soil mechanics as easy and informative as possible, for the readers.

To facilitate a deeper understanding of the contents of this book a short introduction of every chapter is written below:

Chapter 1- Soil is primarily classified according to its use, which can be soil as resource and soil as material. As a resource, soil is studied for its qualities of sustaining agriculture while as a material, soil is studied for geotechnical research. The study has led to classification of soil in different types of ways. This is an introductory chapter which will introduce briefly all the significant aspects of soil mechanics.

Chapter 2- Soil phases relate to the composition of soil which is classified into solids, liquids and gases. Soil analysis calculates porosity, void ratio, air ratio and the degree of saturation with regard to the different soil phases. Soil fertility is the ability of soil to provide a sat-isfactory habitat and productive yield to a plant. It depends on soil's ability to provide nutrients and water, and also on the absence of toxic substances. This chapter has been carefully written to provide an easy understanding of the varied facets of soil mechanics.

Chapter 3- The major components of pressure on land have been discussed. Total vertical stress on a ground surface is the load applied to soil, which gets carried by solid grains and water. Also explored are the topics like soil retrogression and degradation, land degradation, shear strength, etc. This chapter provides a plethora of interdisciplinary topics for better comprehension of soil mechanics.

Chapter 4- Soil contains pores which have gases and liquids present in them. These pores also help in passage of water. Permeability depends on this passage as its value can be determined on the size of pores. Other factors that affect permeability are particle shape, distribution of particle sizes, and soil structure. Also discussed are the concepts of soil contamination, mining space, etc. Soil contamination and permeability are one of the significant and important topics related to soil mechanics. The following chapter unfolds its crucial aspects in a critical yet systematic manner.

Chapter 5- Over usage, erosion and acidification causes loss of fertility in land. Soil conservation involves methods of avoiding erosion and exhaustion. This section helps the readers in developing a better understanding of soil erosion. Soil mechanics is best understood in confluence with the major topics listed in the following chapter.

I would like to share the credit of this book with my editorial team who worked tirelessly on this book. I owe the completion of this book to the never-ending support of my family, who supported me throughout the project.

Editor

1

Understanding Soil Mechanics

Soil is primarily classified according to its use, which can be soil as resource and soil as material. As a resource, soil is studied for its qualities of sustaining agriculture while as a material, soil is studied for geotechnical research. The study has led to classification of soil in different types of ways. This is an introductory chapter which will introduce briefly all the significant aspects of soil mechanics.

Soil Mechanics

Soil mechanics is a branch of soil physics and engineering mechanics that describes the behavior of soils. It differs from fluid mechanics and solid mechanics in the sense that soils consist of a heterogeneous mixture of fluids (usually air and water) and particles (usually clay, silt, sand, and gravel) but soil may also contain organic solids and other matter. Along with rock mechanics, soil mechanics provides the theoretical basis for analysis in geotechnical engineering, a subdiscipline of civil engineering, and engineering geology, a subdiscipline of geology. Soil mechanics is used to analyze the deformations of and flow of fluids within natural and man-made structures that are supported on or made of soil, or structures that are buried in soils. Example applications are building and bridge foundations, retaining walls, dams, and buried pipeline systems. Principles of soil mechanics are also used in related disciplines such as engineering geology, geophysical engineering, coastal engineering, agricultural engineering, hydrology and soil physics.

The Tower of Pisa – an example of a problem due to deformation of soil.

Slope instability issues for a temporary flood control levee in North Dakota, 2009

Earthwork in Germany

This article describes the genesis and composition of soil, the distinction between pore water pressure and inter-granular effective stress, capillary action of fluids in the soil pore spaces, soil classification, seepage and permeability, time dependent change of volume due to squeezing water out of tiny pore spaces, also known as consolidation, shear strength and stiffness of soils. The shear strength of soils is primarily derived from friction between the particles and interlocking, which are very sensitive to the effective stress. The article concludes with some examples of applications of the principles of soil mechanics such as slope stability, lateral earth pressure on retaining walls, and bearing capacity of foundations.

Genesis and composition of Soils

Genesis

The primary mechanism of soil creation is the weathering of rock. All rock types (igneous rock, metamorphic rock and sedimentary rock) may be broken down into small particles to create soil. Weathering mechanisms are physical weathering, chemical weathering, and biological weathering Human activities such as excavation, blasting, and waste disposal, may also create soil. Over geologic time, deeply buried soils may be altered by pressure and temperature to become metamorphic or sedimentary rock, and if melted and solidified again, they would complete the geologic cycle by becoming igneous rock.

Physical weathering includes temperature effects, freeze and thaw of water in cracks, rain, wind, impact and other mechanisms. Chemical weathering includes dissolution of matter composing a rock and precipitation in the form of another mineral. Clay minerals, for example can be formed by weathering of feldspar, which is the most common mineral present in igneous rock.

The most common mineral constituent of silt and sand is quartz, also called silica, which has the chemical name silicon dioxide. The reason that feldspar is most common in rocks but silicon is more prevalent in soils is that feldspar is much more soluble than silica.

Silt, Sand, and Gravel are basically little pieces of broken rocks.

According to the Unified Soil Classification System, silt particle sizes are in the range of 0.002 mm to 0.075 mm and sand particles have sizes in the range of 0.075 mm to 4.75 mm.

Gravel particles are broken pieces of rock in the size range 4.75 mm to 100 mm. Particles larger than gravel are called cobbles and boulders.

Transport

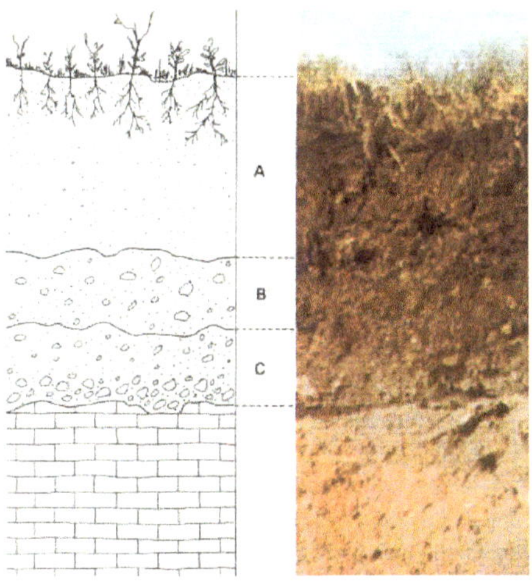

Example soil horizons. a) top soil and colluvium b) mature residual soil c) young residual soil d) weathered rock.

Soil deposits are affected by the mechanism of transport and deposition to their location. Soils that are not transported are called residual soils—they exist at the same location as the rock from which they were generated. Decomposed granite is a common example of a residual soil. The common mechanisms of transport are the actions of gravity, ice, water, and wind. Wind blown soils include dune sands and loess. Water carries particles of different size depending on the speed of the water, thus soils transported by water are graded according to their size. Silt and clay may settle out in a lake, and gravel and sand collect at the bottom of a river bed. Wind blown soil deposits (aeolian soils) also tend to be sorted according to their grain size. Erosion at the base of glaciers is powerful enough to pick up large rocks and boulders as well as soil; soils dropped by melting ice can be a well graded mixture of widely varying particle sizes. Gravity on its own may also carry particles down from the top of a mountain to make a pile of soil and boulders at the base; soil deposits transported by gravity are called colluvium.

The mechanism of transport also has a major effect on the particle shape. For example, low velocity grinding in a river bed will produce rounded particles. Freshly fractured colluvium particles often have a very angular shape.

Soil Composition

Soil Mineralogy

Silts, sands and gravels are classified by their size, and hence they may consist of a variety of minerals. Owing to the stability of quartz compared to other rock minerals, quartz is the most common constituent of sand and silt . Mica, and feldspar are other common minerals present in sands and silts. The mineral constituents of gravel may be more similar to that of the parent rock.

The common clay minerals are montmorillonite or smectite, illite, and kaolinite or kaolin. These minerals tend to form in sheet or plate like structures, with length typically ranging between 10^{-7} m and $4x10^{-6}$ m and thickness typically ranging between 10^{-9} m and $2x10^{-6}$ m, and they have a relatively large specific surface area. The specific surface area (SSA) is defined as the ratio of the surface area of particles to the mass of the particles. Clay minerals typically have specific surface areas in the range of 10 to 1,000 square meters per gram of solid. Due to the large surface area available for chemical, electrostatic, and van der Waals interaction, the mechanical behavior of clay minerals is very sensitive to the amount of pore fluid available and the type and amount of dissolved ions in the pore fluid. To anticipate the effect of clay on the way a soil will behave, it is necessary to know the kinds of clays as well as the amount present. As home builders and highway engineers know all too well, soils containing certain high-activity clays make very unstable material on which to build because they swell when wet and shrink when dry. This shrink-and-swell action can easily crack foundations and cause retaining walls to collapse. These clays also become extremely sticky and difficult to work with when they are wet. In contrast, low-activity clays, formed under different conditions, can be very stable and easy to work with.

The minerals of soils are predominantly formed by atoms of oxygen, silicon, hydrogen, and aluminum, organized in various crystalline forms. These elements along with calcium, sodium, potassium, magnesium, and carbon constitute over 99 per cent of the solid mass of soils.

Grain size distribution

Soils consist of a mixture of particles of different size, shape and mineralogy. Because the size of the particles obviously has a significant effect on the soil behavior, the grain size and grain size distribution are used to classify soils. The grain size distribution describes the relative proportions of particles of various sizes. The grain size is often visualized in a cumulative distribution graph which, for example, plots the percentage of particles finer than a given size as a function of size. The median grain size, D_{50}, is the size for which 50% of the particle mass consists of finer particles. Soil behavior, especially the hydraulic conductivity, tends to be dominated by the smaller particles, hence, the term "effective size", denoted by D_{10}, is defined as the size for which 10% of the particle mass consists of finer particles.

Sands and gravels that possess a wide range of particle sizes with a smooth distribution of particle sizes are called well graded soils. If the soil particles in a sample are predominantly in a relatively narrow range of sizes, the sample is uniformly graded. If a soil sample has distinct gaps in the gradation curve, e.g., a mixture of gravel and fine sand, with no coarse sand, the sample may be gap graded. Uniformly graded and gap graded soils are both considered to be poorly graded. There are many methods for measuring particle size distribution. The two traditional methods are sieve analysis and hydrometer analysis.

Sieve Analysis

Sieve

The size distribution of gravel and sand particles are typically measured using sieve analysis. The formal procedure is described in ASTM D6913-04(2009). A stack of sieves with accurately dimensioned holes between a mesh of wires is used to separate the particles into size bins. A known volume of dried soil, with clods broken down to individual particles, is put into the top of a stack of sieves arranged from coarse to fine. The stack of sieves is shaken for a standard period of time so that the particles are sorted into size bins. This method works reasonably well for particles in the sand and gravel size range. Fine particles tend to stick to each other, and hence the sieving process is not an effective method. If there are a lot of fines (silt and clay) present in the soil it may be necessary to run water through the sieves to wash the coarse particles and clods through.

A variety of sieve sizes are available. The boundary between sand and silt is arbitrary. According to the Unified Soil Classification System, a #4 sieve (4 openings per inch) having 4.75mm opening size separates sand from gravel and a #200 sieve with an 0.075 mm opening separates sand from silt and clay. According to the British standard, 0.063 mm is the boundary between sand and silt, and 2 mm is the boundary between sand and gravel.

Hydrometer Analysis

The classification of fine-grained soils, i.e., soils that are finer than sand, is determined primarily by their Atterberg limits, not by their grain size. If it is important to determine the grain size distribution of fine-grained soils, the hydrometer test may be performed. In the hydrometer tests, the soil particles are mixed with water and shaken to produce a dilute suspension in a glass cylinder, and then the cylinder is left to sit. A hydrometer is used to measure the density of the suspension as a function of time. Clay particles may take several hours to settle past the depth of measurement of the hydrometer. Sand particles may take less than a second. Stoke's law provides the theoretical basis to calculate the relationship between sedimentation velocity and particle size. ASTM provides the detailed procedures for performing the Hydrometer test.

Clay particles can be sufficiently small that they never settle because they are kept in suspension by Brownian motion, in which case they may be classified as colloids.

Mass-volume relations

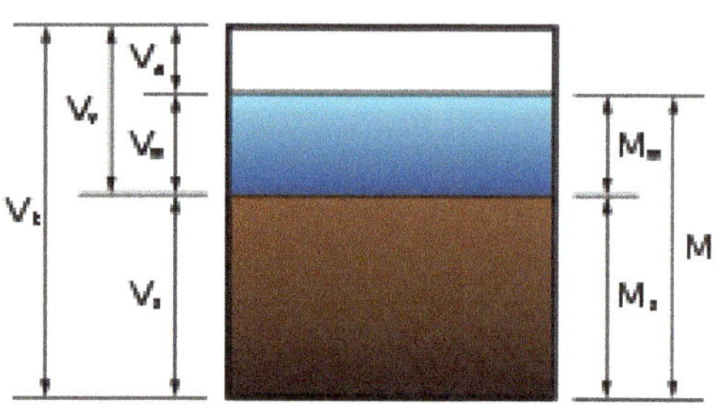

A phase diagram of soil indicating the masses and volumes of air, solid, water, and voids.

There are a variety of parameters used to describe the relative proportions of air, water and solid in a soil. This section defines these parameters and some of their interrelationships. The basic notation is as follows:

V_a, V_w and V_s represent the volumes of air, water and solids in a soil mixture;

W_a, W_w and W_s represent the weights of air, water and solids in a soil mixture;

M_a, M_w and M_s represent the masses of air, water and solids in a soil mixture;

ρ_a, ρ_w, and ρ_s represent the densities of the constituents (air, water and solids) in a soil mixture;

Note that the weights, W, can be obtained by multiplying the mass, M, by the acceleration due to gravity, g; e.g., $W_s = M_s g$.

Specific Gravity is the ratio of the density of one material compared to the density of pure water ($\rho_w = 1 g / cm^3$).

Specific gravity of solids, $G_s = \frac{\rho_s}{\rho_w}$

Note that unit weights, conventionally denoted by the symbol γ may be obtained by multiplying the density (ρ) of a material by the acceleration due to gravity, g.

Density, Bulk Density, or Wet Density, ρ, are different names for the density of the mixture, i.e., the total mass of air, water, solids divided by the total volume of air water and solids (the mass of air is assumed to be zero for practical purposes):

$$\rho = \frac{M_s + M_w}{V_s + V_w + V_a} = \frac{M_t}{V_t}$$

Dry Density, ρ_d, is the mass of solids divided by the total volume of air water and solids:

$$\rho_d = \frac{M_s}{V_s + V_w + V_a} = \frac{M_s}{V_t}$$

Buoyant Density, ρ', defined as the density of the mixture minus the density of water is useful if the soil is submerged under water:

$$\rho' = \rho - \rho_w$$

where ρ_w is the density of water

Water Content, w is the ratio of mass of water to mass of solid. It is easily measured by weighing a sample of the soil, drying it out in an oven and re-weighing. Standard procedures are described by ASTM.

$$w = \frac{M_w}{M_s} = \frac{W_w}{W_s}$$

Void ratio, e, is the ratio of the volume of voids to the volume of solids:

$$e = \frac{V_v}{V_s} = \frac{V_v}{V_T - V_V} = \frac{n}{1-n}$$

Porosity, n, is the ratio of volume of voids to the total volume, and is related to the void ratio:

$$n = \frac{V_v}{V_t} = \frac{V_v}{V_s + V_v} = \frac{e}{1+e}$$

Degree of saturation, S, is the ratio of the volume of water to the volume of voids:

$$S = \frac{V_w}{V_v}$$

From the above definitions, some useful relationships can be derived by use of basic algebra.

$$\rho = \frac{(G_s + Se)\rho_w}{1+e}$$

$$\rho = \frac{(1+w)G_s\rho_w}{1+e}$$

$$w = \frac{Se}{G_s}$$

Soil Classification

Geotechnical engineers classify the soil particle types by performing tests on disturbed (dried, passed through sieves, and remolded) samples of the soil. This provides information about the characteristics of the soil grains themselves. It should be noted that classification of the types of grains present in a soil does not account for important effects of the structure or fabric of the soil, terms that describe compactness of the particles and patterns in the arrangement of particles in a load carrying framework as well as the pore size and pore fluid distributions. Engineering geologists also classify soils based on their genesis and depositional history.

Classification of Soil Grains

In the US and other countries, the Unified Soil Classification System (USCS) is often used for soil classification. Other classification systems include the British Standard BS5390 and the AASHTO soil classification system.

Classification of Sands and Gravels

In the USCS, gravels (given the symbol G) and sands (given the symbol S) are classified according to their grain size distribution. For the USCS, gravels may be given the classification symbol GW (well-graded gravel), GP (poorly graded gravel), GM (gravel with a large amount of silt), or GC (gravel with a large amount of clay). Likewise sands may be classified as being SW, SP, SM or SC. Sands and gravels with a small but non-negligible amount of fines (5–12%) may be given a dual classification such as SW-SC.

Atterberg Limits

Clays and Silts, often called 'fine-grained soils', are classified according to their Atterberg limits; the most commonly used Atterberg limits are the Liquid Limit (denoted by LL or w_l), Plastic Limit (denoted by PL or w_p), and Shrinkage Limit (denoted by SL).

The Liquid Limit is the water content at which the soil behavior transitions from that of a liquid to that of a plastic solid. The Plastic Limit is the water content at which the soil behavior transitions from that of a plastic solid to a brittle solid. The Shrinkage Limit corresponds to a water content below which the soil will not shrink as it dries.

As the transitions from one state to another are gradual, the tests have adopted arbitrary definitions to determine the boundaries of the states. The liquid limit is determined by measuring the water content for which a groove closes after 25 blows in a standard test. Alternatively, a fall cone test apparatus may be use to measure the liquid limit. The undrained shear strength of remolded soil at the liquid limit is approximately 2 kPa. The Plastic Limit is the water content below which it is not possible to roll by hand the soil into 3 mm diameter cylinders. The soil cracks or breaks up as it is rolled down to this diameter. Remolded soil at the plastic limit is quite stiff, having an undrained shear strength of the order of about 200 kPa.

The Plasticity Index of a particular soil specimen is defined as the difference between the Liquid Limit and the Plastic Limit of the specimen; it is an indicator of how much water the soil particles in the specimen can absorb, and correlates with many engineering properties.

Classification of Silts and Clays

According to the Unified Soil Classification System (USCS), silts and clays are classified by plotting the values of their plasticity index and liquid limit on a plasticity chart. The A-Line on the chart separates clays (given the USCS symbol C) from silts (given the symbol M). LL=50% separates high plasticity soils (given the modifier symbol H) from low plasticity soils (given the modifier symbol L). A soil that plots above the A-line and has LL>50% would, for example, be classified as CH. Other possible classifications of silts and clays are ML, CL and MH. If the Atterberg limits plot in the"hatched" region on the graph near the origin, the soils are given the dual classification 'CL-ML'.

Indices Related to Soil Strength

Liquidity Index

The effects of the water content on the strength of saturated remolded soils can be quantified by the use of the liquidity index, LI:

$$LI = \frac{w - PL}{LL - PL}$$

When the LI is 1, remolded soil is at the liquid limit and it has an undrained shear strength of about 2 kPa. When the soil is at the plastic limit, the LI is 0 and the undrained shear strength is about 200 kPa.

Relative Density

The density of sands (cohesionless soils) is often characterized by the relative density, D_r

$$D_r = \frac{e_{max} - e}{e_{max} - e_{min}} 100\%$$

where: e_{max} is the "maximum void ratio" corresponding to a very loose state, e_{min} is the "minimum void ratio" corresponding to a very dense state and e is the in situ void ratio. Methods used to calculate relative density are defined in ASTM D4254-00 (2006).

Thus if $D_r = 100\%$ the sand or gravel is very dense, and if $D_r = 0\%$ the soil is extremely loose and unstable.

Seepage: Steady State Flow of Water

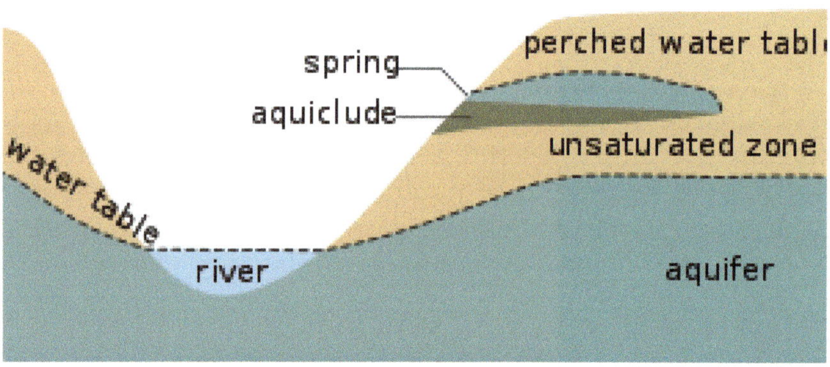

A cross section showing the water table varying with surface topography as well as a perched water table

If fluid pressures in a soil deposit are uniformly increasing with depth according to $u = \rho_w g z_w$ then hydrostatic conditions will prevail and the fluids will not be flowing through the soil. z_w is the depth below the water table. However, if the water table is sloping or there is a perched water table as indicated in the accompanying sketch, then seepage will occur. For steady state seepage, the seepage velocities are not varying with time. If the water tables are changing levels with time, or if the soil is in the process of consolidation, then steady state conditions do not apply.

Darcy's law

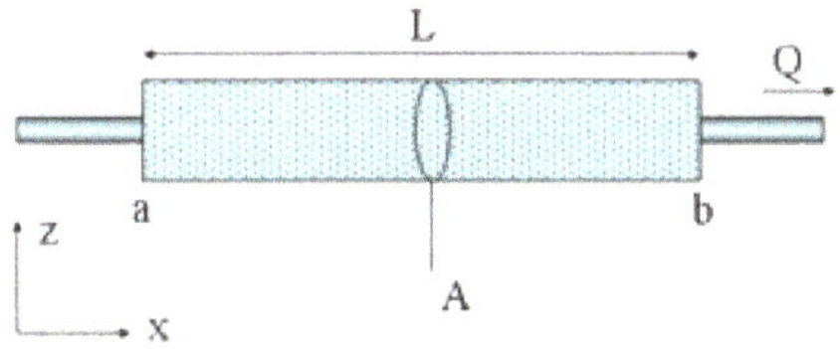

Diagram showing definitions and directions for Darcy's law

Darcy's law states that the volume of flow of the pore fluid through a porous medium per unit time is proportional to the rate of change of excess fluid pressure with distance. The constant of proportionality includes the viscosity of the fluid and the intrinsic permeability of the soil. For the simple case of a horizontal tube filled with soil

$$Q = \frac{-KA}{\mu}\frac{(u_b - u_a)}{L}$$

The total discharge, Q (having units of volume per time, e.g., ft³/s or m³/s), is proportional to the intrinsic permeability, K, the cross sectional area, A, and rate of pore pressure change with distance, $\frac{u_b - u_a}{L}$, and inversely proportional to the dynamic viscosity of the fluid, μ. The negative sign is needed because fluids flow from high pressure to low pressure. So if the change in pressure is negative (in the x-direction) then the flow will be positive (in the x-direction). The above equation works well for a horizontal tube, but if the tube was inclined so that point b was a different elevation than point a, the equation would not work. The effect of elevation is accounted for by replacing the pore pressure by excess pore pressure, u_e defined as:

$$u_e = u - \rho_w gz$$

where z is the depth measured from an arbitrary elevation reference (datum). Replacing u by u_e we obtain a more general equation for flow:

$$Q = \frac{-KA}{\mu}\frac{(u_{e,b} - u_{e,a})}{L}$$

Dividing both sides of the equation by A, and expressing the rate of change of excess pore pressure as a derivative, we obtain a more general equation for the apparent velocity in the x-direction:

$$v_x = \frac{-K}{\mu}\frac{du_e}{dx}$$

where $v_x = Q / A$ has units of velocity and is called the Darcy velocity (or the specific discharge, filtration velocity, or superficial velocity). The pore or interstitial velocity v_{px} is the average velocity of fluid molecules in the pores; it is related to the Darcy velocity and the porosity n through the Dupuit-Forchheimer relationship

$$v_{px} = \frac{v_x}{n}$$

(Some authors use the term seepage velocity to mean the Darcy velocity, while others use it to mean the pore velocity.)

Civil engineers predominantly work on problems that involve water and predominantly work on problems on earth (in earth's gravity). For this class of problems, civil engineers will often write Darcy's law in a much simpler form:

$$v = ki$$

where k is the hydraulic conductivity, defined as $k = \frac{K\rho_w g}{\mu_w}$, and i is the hydraulic gradient. The hydraulic gradient is the rate of change of total head with distance. The total head, h at a point is defined as the height (measured relative to the datum) to which water would rise in a piezometer at that point. The total head is related to the excess water pressure by:

$$u_e = \rho_w gh + Constant$$

and the $Constant$ is zero if the datum for head measurement is chosen at the same elevation as the origin for the depth, z used to calculate u_e.

Typical Values of Hydraulic Conductivity

Values of hydraulic conductivity, k, can vary by many orders of magnitude depending on the soil type. Clays may have hydraulic conductivity as small as about $10^{-12}\frac{m}{s}$, gravels may have hydraulic conductivity up to about $10^{-1}\frac{m}{s}$. Layering and heterogeneity and disturbance during the sampling and testing process make the accurate measurement of soil hydraulic conductivity a very difficult problem.

Flownets

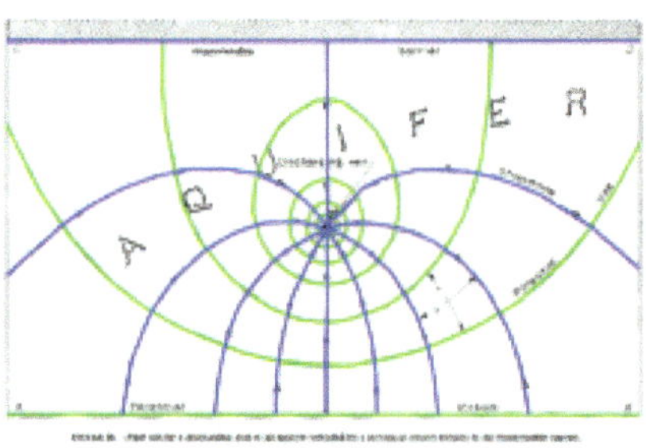

A plan flow net to estimate flow of water from a stream to a discharging well

Darcy's Law applies in one, two or three dimensions. In two or three dimensions, steady state seepage is described by Laplace's equation. Computer programs are available to solve this equation. But traditionally two-dimensional seepage problems were solved using a graphical procedure known as flownet. One set of lines in the flownet are in the direction of the water flow (flow lines), and the other set of lines are in the direction of constant total head (equipotential lines). Flownets may be used to estimate the quantity of seepage under dams and sheet piling.

Seepage Forces and Erosion

When the seepage velocity is great enough, erosion can occur because of the frictional drag exerted on the soil particles. Vertically upwards seepage is a source of danger on the downstream side of sheet piling and beneath the toe of a dam or levee. Erosion of the soil, known as "soil piping", can lead to failure of the structure and to sinkhole formation. Seeping water removes soil, starting from the exit point of the seepage, and erosion advances upgradient. The term "sand boil" is used to describe the appearance of the discharging end of an active soil pipe.

Seepage Pressures

Seepage in an upward direction reduces the effective stress within the soil. When the water pressure at a point in the soil is equal to the total vertical stress at that point, the effective stress is zero and the soil has no frictional resistance to deformation. For a surface layer, the vertical effective stress becomes zero within the layer when the upward hydraulic gradient is equal to the critical gradient. At zero effective stress soil has very little strength and layers of relatively impermeable soil may heave up due to the underlying water pressures. The loss in strength due to upward seepage is a common contributor to levee failures. The condition of zero effective stress associated with upward seepage is also called liquefaction, quicksand, or a boiling condition. Quicksand was so named because the soil particles move around and appear to be 'alive' (the biblical meaning of 'quick' – as opposed to 'dead'). (Note that it is not possible to be 'sucked down' into quicksand. On the contrary, you would float with about half your body out of the water.)

Effective Stress and Capillarity: Hydrostatic Conditions

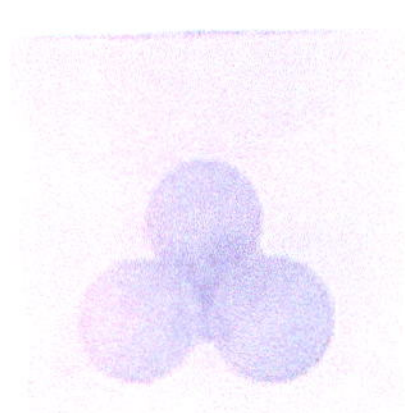

Spheres immersed in water, reducing effective stress.

To understand the mechanics of soils it is necessary to understand how normal stresses and shear stresses are shared by the different phases. Neither gas nor liquid provide significant resistance to shear stress. The shear resistance of soil is provided by friction and interlocking of the particles. The friction depends on the intergranular contact stresses between solid particles. The normal stresses, on the other hand, are shared by the fluid and the particles. Although the pore air is relatively compressible, and hence takes little normal stress in most geotechnical problems, liquid water is relatively incompressible and if the voids are saturated with water, the pore water must be squeezed out in order to pack the particles closer together.

The principle of effective stress, introduced by Karl Terzaghi, states that the effective stress σ› (i.e., the average intergranular stress between solid particles) may be calculated by a simple subtraction of the pore pressure from the total stress:

$$\sigma' = \sigma - u$$

where σ is the total stress and u is the pore pressure. It is not practical to measure σ› directly, so in practice the vertical effective stress is calculated from the pore pressure and vertical total stress. The distinction between the terms pressure and stress is also important. By definition, pressure at a point is equal in all directions but stresses at a point can be different in different directions. In soil mechanics, compressive stresses and pressures are considered to be positive and tensile stresses are considered to be negative, which is different from the solid mechanics sign convention for stress.

Total Stress

For level ground conditions, the total vertical stress at a point, σ_v, on average, is the weight of everything above that point per unit area. The vertical stress beneath a uniform surface layer with density ρ, and thickness H is for example:

$$\sigma_v = \rho g H = \gamma H$$

where g is the acceleration due to gravity, and γ is the unit weight of the overlying layer. If there are multiple layers of soil or water above the point of interest, the vertical stress may be calculated by summing the product of the unit weight and thickness of all of the overlying layers. Total stress increases with increasing depth in proportion to the density of the overlying soil.

It is not possible to calculate the horizontal total stress in this way. Lateral earth pressures are addressed elsewhere.

Hydrostatic Conditions

If there is no pore water flow occurring in the soil, the pore water pressures will be hydrostatic. The water table is located at the depth where the water pressure is equal to the atmospheric pressure. For hydrostatic conditions, the water pressure increases linearly with depth below the water table:

$$u = \rho_w g z_w$$

where ρ_w is the density of water, and z_w is the depth below the water table.

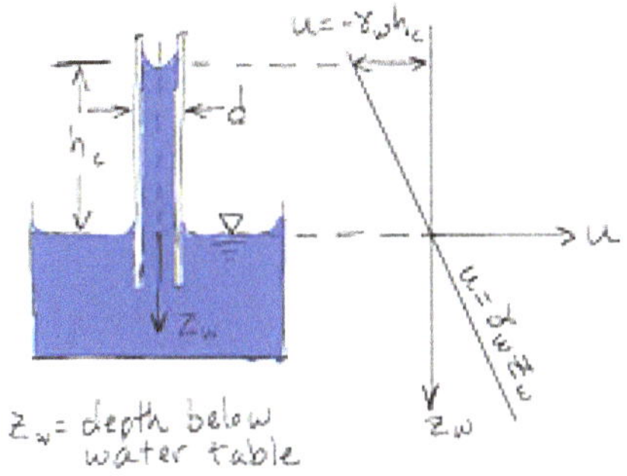

Water is drawn into a small tube by surface tension. Water pressure, u, is negative above and positive below the free water surface

Capillary Action

Due to surface tension, water will rise up in a small capillary tube above a free surface of water. Likewise, water will rise up above the water table into the small pore spaces around the soil particles. In fact the soil may be completely saturated for some distance above the water table. Above the height of capillary saturation, the soil may be wet but the water content will decrease with elevation. If the water in the capillary zone is not moving, the water pressure obeys the equation of hydrostatic equilibrium, $u = \rho_w g z_w$, but note that z_w, is negative above the water table. Hence, hydrostatic water pressures are negative above the water table. The thickness of the zone of capillary saturation depends on the pore size, but typically, the heights vary between a centimeter or so for coarse sand to tens of meters for a silt or clay. In fact the pore space of soil is a uniform fractal e.g. a set of uniformly distributed D-dimensional fractals of average linear size L. For the clay soil it has been found that L=0.15 mm and D=2.7.

The surface tension of water explains why the water does not drain out of a wet sand castle or a moist ball of clay. Negative water pressures make the water stick to the particles and pull the particles to each other, friction at the particle contacts make a sand castle stable. But as soon as a wet sand castle is submerged below a free water surface, the negative pressures are lost and the castle collapses. Considering the effective stress equation, $\sigma' = \sigma - u$, if the water pressure is negative, the effective stress may be positive, even on a free surface (a surface where the total normal stress is zero). The negative pore pressure pulls the particles together and causes compressive particle to particle contact forces. Negative pore pressures in clayey soil can be much more powerful than those in sand. Negative pore pressures explain why clay soils shrink when they dry and swell as they are wetted. The swelling and shrinkage can cause major distress, especially to light structures and roads.

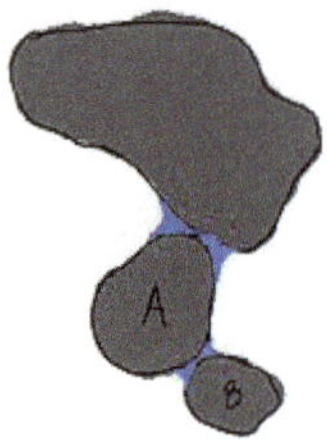

Water at particle contacts

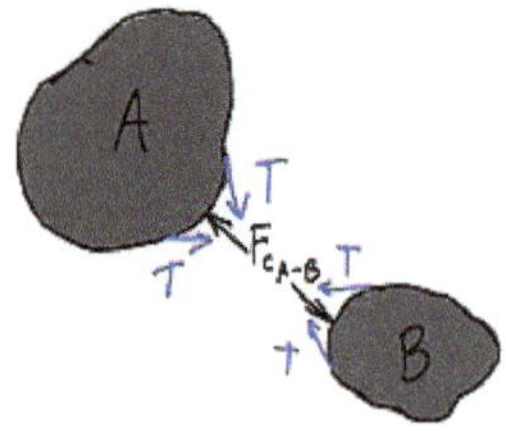

intergranular contact force due to surface tension

Shrinkage caused by drying

Consolidation: Transient Flow of Water

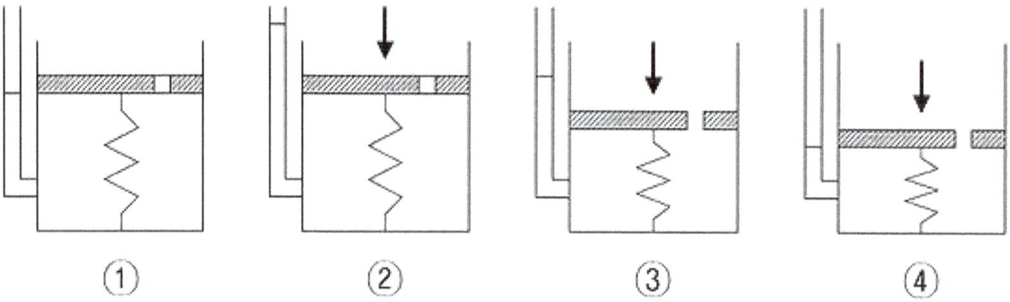

Consolidation analogy. The piston is supported by water underneath and a spring. When a load is applied to the piston, water pressure increases to support the load. As the water slowly leaks through the small hole, the load is transferred from the water pressure to the spring force.

Consolidation is a process by which soils decrease in volume. It occurs when stress is applied to a soil that causes the soil particles to pack together more tightly, therefore reducing volume. When this occurs in a soil that is saturated with water, water will be squeezed out of the soil. The time required to squeeze the water out of a thick deposit of clayey soil layer might be years. For a layer of sand, the water may be squeezed out in a matter of seconds. A building foundation or construction of a new embankment will cause the soil below to consolidate and this will cause settlement which in turn may cause distress to the building or embankment. Karl Terzaghi developed the theory of consolidation which enables prediction of the amount of settlement and the time required for the settlement to occur. Soils are tested with an oedometer test to determine their compression index and coefficient of consolidation.

When stress is removed from a consolidated soil, the soil will rebound, drawing water back into the pores and regaining some of the volume it had lost in the consolidation process. If the stress is reapplied, the soil will re-consolidate again along a recompression curve, defined by the recompression index. Soil that has been consolidated to a large pressure and has been subsequently unloaded is considered to be overconsolidated. The maximum past vertical effective stress is termed the preconsolidation stress. A soil which is currently experiencing the maximum past vertical effective stress is said to be normally consolidated. The overconsolidation ratio, (OCR) is the ratio of the maximum past vertical effective stress to the current vertical effective stress. The OCR is significant for two reasons: firstly, because the compressibility of normally consolidated soil is significantly larger than that for overconsolidated soil, and secondly, the shear behavior and dilatancy of clayey soil are related to the OCR through critical state soil mechanics; highly overconsolidated clayey soils are dilatant, while normally consolidated soils tend to be contractive.

Shear Behavior: Stiffness and Strength

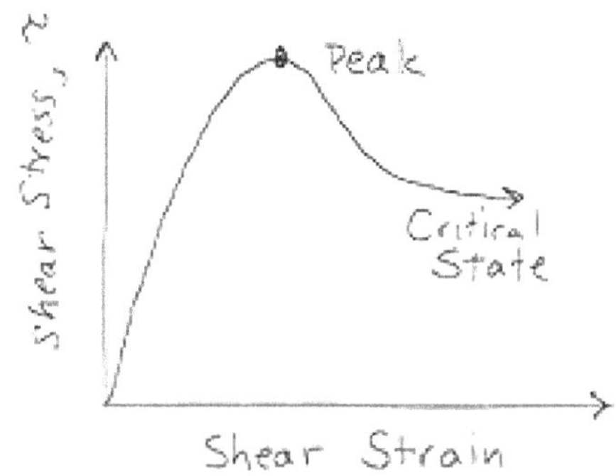

Typical stress strain curve for a drained dilatant soil

The shear strength and stiffness of soil determines whether or not soil will be stable or how much it will deform. Knowledge of the strength is necessary to determine if a slope will be stable, if a building or bridge might settle too far into the ground, and the limiting pressures on a retaining wall. It is important to distinguish between failure of a soil element and the failure of a geotechnical structure (e.g., a building foundation, slope or retaining wall); some soil elements may reach their peak strength prior to failure of the structure. Different criteria can be used to define the "shear strength" and the "yield point" for a soil element from a stress–strain curve. One may define the peak shear strength as the peak of a stress–strain curve, or the shear strength at critical state as the value after large strains when the shear resistance levels off. If the stress–strain curve does not stabilize before the end of shear strength test, the "strength" is sometimes considered to be the shear resistance at 15–20% strain. The shear strength of soil depends on many factors including the effective stress and the void ratio.

The shear stiffness is important, for example, for evaluation of the magnitude of deformations of foundations and slopes prior to failure and because it is related to the shear wave velocity. The slope of the initial, nearly linear, portion of a plot of shear stress as a function of shear strain is called the shear modulus

Friction, Interlocking and Dilation

Soil is an assemblage of particles that have little to no cementation while rock (such as sandstone) may consist of an assembly of particles that are strongly cemented together by chemical bonds. The shear strength of soil is primarily due to interparticle friction and therefore, the shear resistance on a plane is approximately proportional to the effective normal stress on that plane. The angle of internal friction is thus closely related to the maximum stable slope angle, often called the angle of repose.

Angle of repose

But in addition to friction, soil derives significant shear resistance from interlocking of grains. If the grains are densely packed, the grains tend to spread apart from each other as they are subject to shear strain. The expansion of the particle matrix due to shearing was called dilatancy by Osborne Reynolds. If one considers the energy required to shear an assembly of particles there is energy input by the shear force, T, moving a distance, x and there is also energy input by the normal force, N, as the sample expands a distance, y. Due to the extra energy required for the particles to dilate against the confining pressures, dilatant soils have a greater peak strength than contractive soils. Furthermore, as dilative soil grains dilate, they become looser (their void ratio increases), and their rate of dilation decreases until they reach a critical void ratio. Contractive soils become denser as they shear, and their rate of contraction decreases until they reach a critical void ratio.

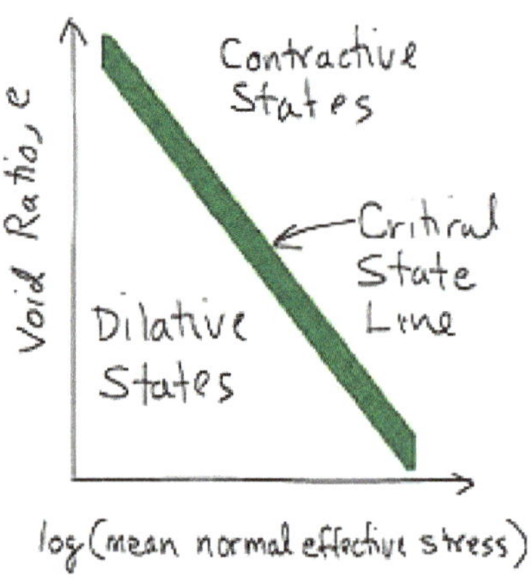

A critical state line separates the dilatant and contractive states for soil

The tendency for a soil to dilate or contract depends primarily on the confining pressure and the void ratio of the soil. The rate of dilation is high if the confining pressure is small and the void ratio is small. The rate of contraction is high if the confining pressure is large and the void ratio is large. As a first approximation, the regions of contraction and dilation are separated by the critical state line.

Failure Criteria

After a soil reaches the critical state, it is no longer contracting or dilating and the shear stress on the failure plane $\sigma_{n'}$ is determined by the effective normal stress on the failure plane $\sigma_{n'}$ and critical state friction angle $\phi_{crit'}$:

$$\tau_{crit} = \sigma_{n'} \tan \phi_{crit}$$

The peak strength of the soil may be greater, however, due to the interlocking (dilatancy) contribution. This may be stated:

$$\tau_{peak} = \sigma_{n'} \tan \phi_{peak}$$

Where $\phi_{peak'} > \phi_{crit'}$. However, use of a friction angle greater than the critical state value for design requires care. The peak strength will not be mobilized everywhere at the same time in a practical problem such as a foundation, slope or retaining wall. The critical state friction angle is not nearly as variable as the peak friction angle and hence it can be relied upon with confidence.

Not recognizing the significance of dilatancy, Coulomb proposed that the shear strength of soil may be expressed as a combination of adhesion and friction components:

$$\tau_f = c' + \sigma_{f'} \tan \phi'$$

It is now known that the c' and ϕ' parameters in the last equation are not fundamental soil properties. In particular, c' and ϕ' are different depending on the magnitude of effective stress. According to Schofield (2006), the longstanding use of c' in practice has led many engineers to wrongly believe that c' is a fundamental parameter. This assumption that c' and ϕ' are constant can lead to overestimation of peak strengths.

Structure, fabric, and chemistry

In addition to the friction and interlocking (dilatancy) components of strength, the structure and fabric also play a significant role in the soil behavior. The structure and fabric include factors such as the spacing and arrangement of the solid particles or the amount and spatial distribution of pore water; in some cases cementitious material accumulates at particle-particle contacts. Mechanical behavior of soil is affected by the density of the particles and their structure or arrangement of the particles as well as the amount and spatial distribution of fluids present (e.g., water and air voids). Other factors include the electrical charge of the particles, chemistry of pore water, chemical bonds (i.e. cementation -particles connected through a solid substance such as recrystallized calcium carbonate)

Drained and Undrained Shear

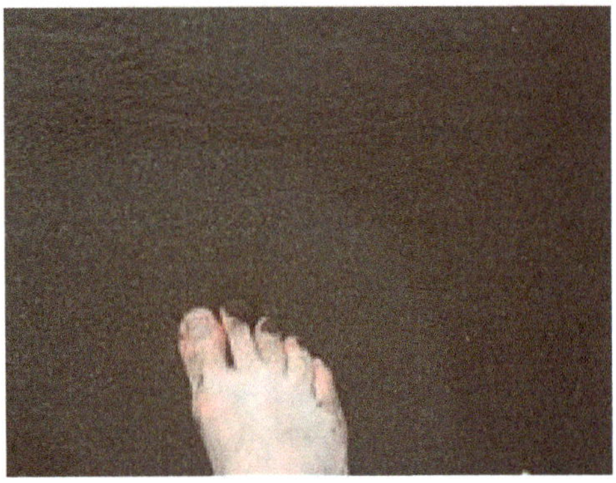

Foot pressure on sand causes it to dilate, drawing water from the surface into the pores

The presence of nearly incompressible fluids such as water in the pore spaces affects the ability for the pores to dilate or contract.

If the pores are saturated with water, water must be sucked into the dilating pore spaces to fill the expanding pores (this phenomenon is visible at the beach when apparently dry spots form around feet that press into the wet sand).

Similarly, for contractive soil, water must be squeezed out of the pore spaces to allow contraction to take place.

Dilation of the voids causes negative water pressures that draw fluid into the pores, and contraction of the voids causes positive pore pressures to push the water out of the pores. If the rate of shearing is very large compared to the rate that water can be sucked into or squeezed out of the dilating or contracting pore spaces, then the shearing is called undrained shear, if the shearing is slow enough that the water pressures are negligible, the shearing is called drained shear. During undrained shear, the water pressure u changes depending on volume change tendencies. From the effective stress equation, the change in u directly effects the effective stress by the equation:

$$\sigma' = \sigma - u$$

and the strength is very sensitive to the effective stress. It follows then that the undrained shear strength of a soil may be smaller or larger than the drained shear strength depending upon whether the soil is contractive or dilative.

Shear Tests

Strength parameters can be measured in the laboratory using direct shear test, triaxial shear test, simple shear test, fall cone test and (hand) shear vane test; there are numerous other devices and variations on these devices used in practice today. Tests conducted to characterize the strength and stiffness of the soils in the ground include the Cone penetration test and the Standard penetration test.

Other factors

The stress–strain relationship of soils, and therefore the shearing strength, is affected by:

1. soil composition (basic soil material): mineralogy, grain size and grain size distribution, shape of particles, pore fluid type and content, ions on grain and in pore fluid.
2. state (initial): Define by the initial void ratio, effective normal stress and shear stress (stress history). State can be describe by terms such as; loose, dense, overconsolidated, normally consolidated, stiff, soft, contractive, dilative, etc.
3. structure: Refers to the arrangement of particles within the soil mass; the manner in which the particles are packed or distributed. Features such as layers, joints, fissures, slickensides, voids, pockets, cementation, etc., are part of the structure. Structure of soils is described by terms such as; undisturbed, disturbed, remolded, compacted, cemented; flocculent, honey-combed, single-grained; flocculated, deflocculated; stratified, layered, laminated; isotropic and anisotropic.
4. Loading conditions: Effective stress path -drained, undrained, and type of loading -magnitude, rate (static, dynamic), and time history (monotonic, cyclic).

Applications

Lateral Earth Pressure

Lateral earth stress theory is used to estimate the amount of stress soil can exert perpendicular to gravity. This is the stress exerted on retaining walls. A lateral earth stress coefficient, K, is defined as the ratio of lateral (horizontal) effective stress to vertical effective stress for cohesionless soils ($K=\sigma›_h/\sigma›_v$). There are three coefficients: at-rest, active, and passive. At-rest stress is the lateral stress in the ground before any disturbance takes place. The active stress state is reached when a wall moves away from the soil under the influence of lateral stress, and results from shear failure due to reduction of lateral stress. The passive stress state is reached when a wall is pushed into the soil far enough to cause shear failure within the mass due to increase of lateral stress. There are many theories for estimating lateral earth stress; some are empirically based, and some are analytically derived.

Bearing Capacity

The bearing capacity of soil is the average contact stress between a foundation and the soil which will cause shear failure in the soil. Allowable bearing stress is the bearing capacity divided by a factor of safety. Sometimes, on soft soil sites, large settlements may occur under loaded foundations without actual shear failure occurring; in such cases, the allowable bearing stress is determined with regard to the maximum allowable settlement. It is important during construction and design stage of a project to evaluate the subgrade strength. The California Bearing Ratio (CBR) test is commonly used to determine the suitability of a soil as a subgrade for design and construction. The field Plate Load Test is commonly used to predict the deformations and failure characteristics of the soil/subgrade and modulus of subgrade reaction (ks). The Modulus of subgrade reaction (ks) is used in foundation design, soil-structure interaction studies and design of highway pavements.

Slope Stability

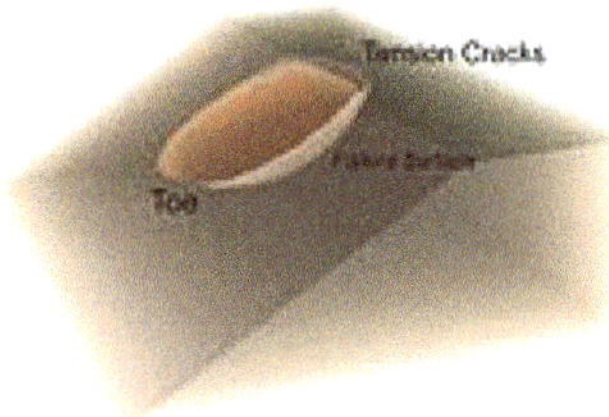

Simple slope slip section

The field of slope stability encompasses the analysis of static and dynamic stability of slopes of earth and rock-fill dams, slopes of other types of embankments, excavated slopes, and natural slopes in soil and soft rock.

As seen to the right, earthen slopes can develop a cut-spherical weakness zone. The probability of this happening can be calculated in advance using a simple 2-D circular analysis package... A primary difficulty with analysis is locating the most-probable slip plane for any given situation. Many landslides have been analyzed only after the fact.

Recent Developments

A recent finding in soil mechanics is that soil deformation can be described as the behavior of a dynamical system. This approach to soil mechanics is referred to as Dynamical Systems based Soil Mechanics (DSSM). DSSM holds simply that soil deformation is a Poisson process in which particles move to their final position at random shear strains.

The basis of DSSM is that soils (including sands) can be sheared till they reach a steady-state condition at which, under conditions of constant strain-rate, there is no change in shear stress, effective confining stress, and void ratio. The steady-state was formally defined by Steve J. Poulos an Associate Professor at the Soil Mechanics Department of Harvard University, who built off a hypothesis that Arthur Casagrande was formulating towards the end of his career. The steady state condition is not the same as the "critical state" condition.It differs from the critical state in that it specifies a statistically constant structure at the steady state. The steady-state values are also very slightly dependent on the strain-rate.

Many systems in nature reach steady-states and dynamical systems theory is used to describe such systems. Soil shear can also be described as a dynamical system. The physical basis of the soil shear dynamical system is a Poisson process in which particles move to the steady-state at random shear strains. Joseph generalized this—particles move to their final position (not just steady-state) at random shear-strains. Because of its origins in the steady state concept DSSM is sometimes informally called "Harvard soil mechanics."

DSSM provides for very close fits to stress–strain curves, including for sands. Because it tracks conditions on the failure plane, it also provides close fits for the post failure region of sensitive clays and silts something that other theories are not able to do. Additionally DSSM explains key relationships in soil mechanics that to date have simply been taken for granted, for example, why normalized undrained peak shear strengths vary with the log of the over consolidation ratio and why stress–strain curves normalize with the initial effective confining stress; and why in one-dimensional consolidation the void ratio must vary with the log of the effective vertical stress, why the end-of-primary curve is unique for static load increments, and why the ratio of the creep value Cα to the compression index Cc must be approximately constant for a wide range of soils.

Soil Classification

Soil classification deals with the systematic categorization of soils based on distinguishing characteristics as well as criteria that dictate choices in use.

Soil types

Overview

Soil classification is a dynamic subject, from the structure of the system itself, to the definitions of classes, and finally in the application in the field. Soil classification can be approached from the perspective of soil as a material and soil as a resource.

Engineering

Engineers, typically geotechnical engineers, classify soils according to their engineering properties as they relate to use for foundation support or building material. Modern engineering classification systems are designed to allow an easy transition from field observations to basic predictions of soil engineering properties and behaviors.

The most common engineering classification system for soils in North America is the Unified Soil Classification System (USCS). The USCS has three major classification groups: (1) coarse-grained soils (e.g. sands and gravels); (2) fine-grained soils (e.g. silts and clays); and (3) highly organic soils (referred to as "peat"). The USCS further subdivides the three major soil classes for clarification. It distinguishes sands from gravels by grain size, and further classifying some as "well-graded" and the rest as "poorly-graded". Silts and clays are distinguished by the soils' Atterberg limits, and separates "high-plasticity" from "low-plasticity" soils as well. Moderately organic soils are considered subdivisions of silts and clays, and are distinguished from inorganic soils by changes in their plasticity properties (and Atterberg limits) on drying. The European soil classification system (ISO 14688) is very similar, differing primarily in coding and in adding an "intermediate-plasticity" classification for silts and clays, and in minor details.

Other engineering soil classification systems in the United States include the AASHTO Soil Classification System, which classifies soils and aggregates relative to their suitability for pavement construction, and the Modified Burmister system, which works similarly to the USCS, but includes more coding for various soil properties.

A full geotechnical engineering soil description will also include other properties of the soil including color, in-situ moisture content, in-situ strength, and somewhat more detail about the material properties of the soil than is provided by the USCS code. The USCS and additional engineering description is standardized in ASTM D 2487.

Soil Science

For soil resources, experience has shown that a natural system approach to classification, i.e. grouping soils by their intrinsic property (soil morphology), behaviour, or genesis, results in classes that can be interpreted for many diverse uses. Differing concepts of pedogenesis, and differences in the significance of morphological features to various land uses can affect the classification approach. Despite these differences, in a well-constructed system, classification criteria group similar concepts so that interpretations do not vary widely. This is in contrast to a technical system approach to soil classification, where soils are grouped according to their fitness for a specific use and their edaphic characteristics.

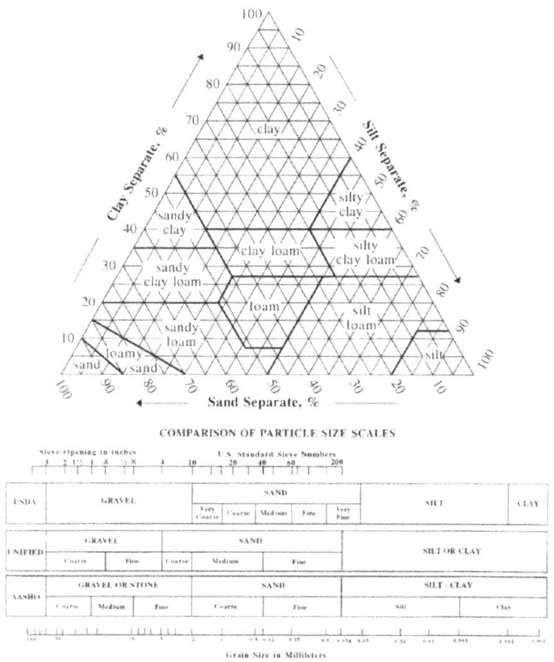

Soil texture triangle showing the USDA classification system based on grain size

Natural system approaches to soil classification, such as the French Soil Reference System (Référentiel pédologique français) are based on presumed soil genesis. Systems have developed, such as USDA soil taxonomy and the World Reference Base for Soil Resources, which use taxonomic criteria involving soil morphology and laboratory tests to inform and refine hierarchical classes.

Another approach is numerical classification, also called ordination, where soil individuals are grouped by multivariate statistical methods such as cluster analysis. This produces natural groupings without requiring any inference about soil genesis.

In soil survey, as practiced in the United States, soil classification usually means criteria based on soil morphology in addition to characteristics developed during soil formation. Criteria are designed to guide choices in land use and soil management. As indicated, this is a hierarchical system that is a hybrid of both *natural* and objective criteria. USDA soil taxonomy provides the core criteria for differentiating soil map units. This is a substantial revision of the 1938 USDA soil taxonomy which was a strictly natural system. Soil taxonomy based soil map units are additionally sorted into classes based on technical classification systems. Land Capability Classes, hydric soil, and prime farmland are some examples.

In addition to scientific soil classification systems, there are also vernacular soil classification systems. Folk taxonomies have been used for millennia, while scientifically based systems are relatively recent developments.

Soil Classifications for OSHA

The U.S. Occupational Safety and Health Administration (OSHA) requires the classification of soils to protect workers from injury when working in excavations and trenches. OSHA uses 3 soil classifications plus one for rock, based primarily on strength but also :

- Stable Rock: natural solid mineral matter that can be excavated with vertical sides and remain intact while exposed.
- Type A - cohesive, plastic soils with unconfined compressive strength greater than 1.5 ton per square foot (tsf)(144 kPa), and meeting several other requirements (with a lateral soil pressure of 25 psf per ft of depth)
- Type B - cohesive soils with unconfined compressive strength between 0.5 tsf (48 kPa) and 1.5 tsf (144 kPa), or unstable dry rock, or soils which would otherwise be Type A (with a lateral soil pressure of 45 psf per ft of depth)
- Type C - granular soils or cohesive soils with unconfined compressive strength less than 0.5 tsf (48 kPa) or any submerged or freely seeping soil or adversely bedded soils (with a lateral soil pressure of 80 psf per ft of depth)
- Type C60 - A subtype of Type C soil, though is not officially recognized by OSHA as a separate type, has a lateral soil pressure of 60 psf per ft of depth

Each of the soil classifications has implications for the way the excavation must be made or the protections (sloping, shoring, shielding, etc.) that must be provided to protect workers from collapse of the excavated bank.

Particle Size Distribution

For measuring the distribution of particle sizes in a soil sample, it is necessary to conduct different particle-size tests.

Wet sieving is carried out for separating fine grains from coarse grains by washing the soil specimen on a 75 micron sieve mesh.

Dry sieve analysis is carried out on particles coarser than 75 micron. Samples (with fines removed) are dried and shaken through a set of sieves of descending size. The weight retained in each sieve is measured. The cumulative percentage quantities finer than the sieve sizes (passing each given sieve size) are then determined.

The resulting data is presented as a distribution curve with grain size along x-axis (log scale) and percentage passing along y-axis (arithmetic scale).

Sedimentation analysis is used only for the soil fraction finer than 75 microns. Soil particles are allowed to settle from a suspension. The decreasing density of the suspension is measured at various time intervals. The procedure is based on the principle that in a suspension, the terminal velocity of a spherical particle is governed by the diameter of the particle and the properties of the suspension.

In this method, the soil is placed as a suspension in a jar filled with distilled water to which a deflocculating agent is added. The soil particles are then allowed to settle down. The concentration of particles remaining in the suspension at a particular level can be determined by using a hydrometer. Specific gravity readings of the solution at that same level at different time intervals provide information about the size of particles that have settled down and the mass of soil remaining in solution.

The results are then plotted between % finer (passing) and log size.

Grain-Size Distribution Curve

The size distribution curves, as obtained from coarse and fine grained portions, can be combined to form one complete grain-size distribution curve (also known as grading curve). A typical grading curve is shown.

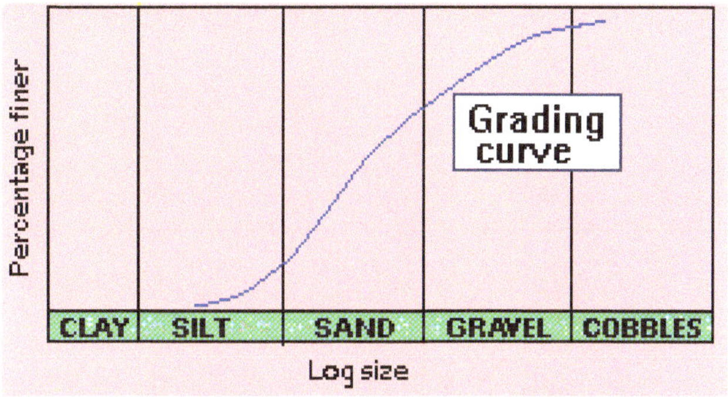

From the complete grain-size distribution curve, useful information can be obtained such as:

1. Grading characteristics, which indicate the uniformity and range in grain-size distribution.

2. Percentages (or fractions) of gravel, sand, silt and clay-size.

Grain Size

Particle size, also called grain size, refers to the diameter of individual grains of sediment, or the lithified particles in clastic rocks. The term may also be applied to other granular materials. This is different from the crystallite size, which refers to the size of a single crystal inside a particle or grain. A single grain can be composed of several crystals. Granular material can range from very small colloidal particles, through clay, silt, sand, gravel, and cobbles, to boulders.

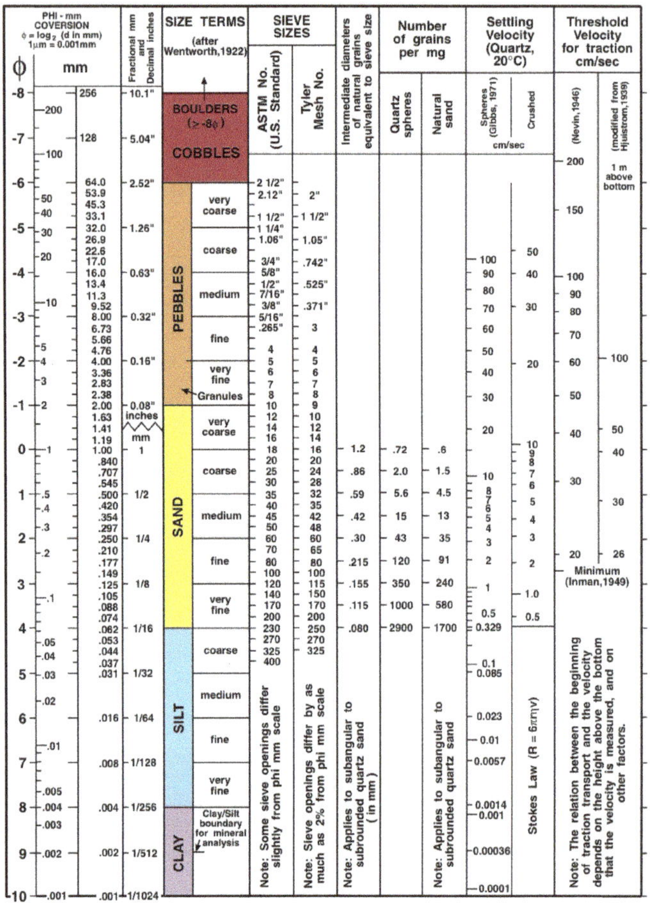

Wentworth grain size chart from United States Geological Survey Open-File Report 2006-1195

Beach cobbles at Nash Point, South Wales.

Krumbein Phi Scale

Size ranges define limits of classes that are given names in the Wentworth scale (or Udden–Wentworth scale) used in the United States. The Krumbein *phi* (φ) scale, a modification of the Wentworth scale created by W. C. Krumbein in 1937, is a logarithmic scale computed by the equation

$$\phi = -\log_2 D / D_0,$$

where

ϕ is the Krumbein phi scale,

D is the diameter of the particle or grain in millimeters (from petrowiki, krumbein and monks equation) and

D_0 is a reference diameter, equal to 1 mm (to make the equation dimensionally consistent).

This equation can be rearranged to find diameter using φ:

$$D = D_0 \times 2^{-\phi}$$

φ scale	Size range (metric)	Size range (approx. inches)	Aggregate name (Wentworth class)	Other names
<−8	>256 mm	>10.1 in	Boulder	
−6 to −8	64–256 mm	2.5–10.1 in	Cobble	
−5 to −6	32–64 mm	1.26–2.5 in	Very coarse gravel	Pebble
−4 to −5	16–32 mm	0.63–1.26 in	Coarse gravel	Pebble
−3 to −4	8–16 mm	0.31–0.63 in	Medium gravel	Pebble
−2 to −3	4–8 mm	0.157–0.31 in	Fine gravel	Pebble
−1 to −2	2–4 mm	0.079–0.157 in	Very fine gravel	Granule
0 to −1	1–2 mm	0.039–0.079 in	Very coarse sand	
1 to 0	0.5–1 mm	0.020–0.039 in	Coarse sand	
2 to 1	0.25–0.5 mm	0.010–0.020 in	Medium sand	
3 to 2	125–250 μm	0.0049–0.010 in	Fine sand	
4 to 3	62.5–125 μm	0.0025–0.0049 in	Very fine sand	
8 to 4	3.9–62.5 μm	0.00015–0.0025 in	Silt	Mud
10 to 8	0.98–3.9 μm	3.8×10^{-5}–0.00015 in	Clay	Mud
20 to 10	0.95–977 nm	3.8×10^{-8}–3.8×10^{-5} in	Colloid	Mud

In some schemes, gravel is anything larger than sand (comprising granule, pebble, cobble, and boulder in the table above).

International Scale

ISO 14688-1:2002, establishes the basic principles for the identification and classification of soils on the basis of those material and mass characteristics most commonly used for soils for engineering purposes. ISO 14688-1 is applicable to natural soils *in situ*, similar man-made materials *in situ* and soils redeposited by people.

ISO 14688-1:2002					
Name				**Size range (mm)**	**Size range (approx. in)**
Very coarse soil		Large boulder	LBo	>630	>24.8031
Boulder		Bo	200–630	7.8740–24.803	
Cobble		Co	63–200	2.4803–7.8740	
Coarse soil	Gravel	Coarse gravel	CGr	20–63	0.78740–2.4803
		Medium gravel	MGr	6.3–20	0.24803–0.78740
		Fine gravel	FGr	2.0–6.3	0.078740–0.24803
	Sand	Coarse sand	CSa	0.63–2.0	0.024803–0.078740
		Medium sand	MSa	0.2–0.63	0.0078740–0.024803
		Fine sand	FSa	0.063–0.2	0.0024803–0.0078740
Fine soil	Silt	Coarse silt	CSi	0.02–0.063	0.00078740–0.0024803
		Medium silt	MSi	0.0063–0.02	0.00024803–0.00078740
		Fine silt	FSi	0.002–0.0063	0.000078740–0.00024803
	Clay		Cl	≤0.002	≤0.000078740

Sorting

An accumulation of sediment can also be characterized by the grain size distribution. A sediment deposit can undergo sorting when a particle size range is removed by an agency such as a river or the wind. The sorting can be quantified using the Inclusive Graphic Standard Deviation:

$$\sigma_I = \frac{\phi 84 - \phi 16}{4} + \frac{\phi 95 - \phi 5}{6.6}$$

where

σ_I is the Inclusive Graphic Standard Deviation in phi units

$\phi 84$ is the 84th percentile of the grain size distribution in phi units, etc.

The result of this can be described using the following terms:

$\sigma_I < 0.35$	$0.35 < \sigma_I < 0.50$	$0.50 < \sigma_I < 1.00$	$1.00 < \sigma_I < 2.00$	$2.00 < \sigma_I < 4.00$	$4.00 < \sigma_I$
very well sorted	well sorted	moderately sorted	poorly sorted	very poorly sorted	extremely poorly sorted

Sieve Analysis

A sieve analysis (or gradation test) is a practice or procedure used (commonly used in civil engineering) to assess the particle size distribution (also called *gradation*) of a granular material.

The size distribution is often of critical importance to the way the material performs in use. A sieve analysis can be performed on any type of non-organic or organic granular materials including sands, crushed rock, clays, granite, feldspars, coal, soil, a wide range of manufactured powders, grain and seeds, down to a minimum size depending on the exact method. Being such a simple technique of particle sizing, it is probably the most common.

Procedure

Sieves used for gradation test.

A gradation test is performed on a sample of aggregate in a laboratory. A typical sieve analysis involves a nested column of sieves with wire mesh cloth (screen).

A representative weighed sample is poured into the top sieve which has the largest screen openings. Each lower sieve in the column has smaller openings than the one above. At the base is a round pan, called the receiver.

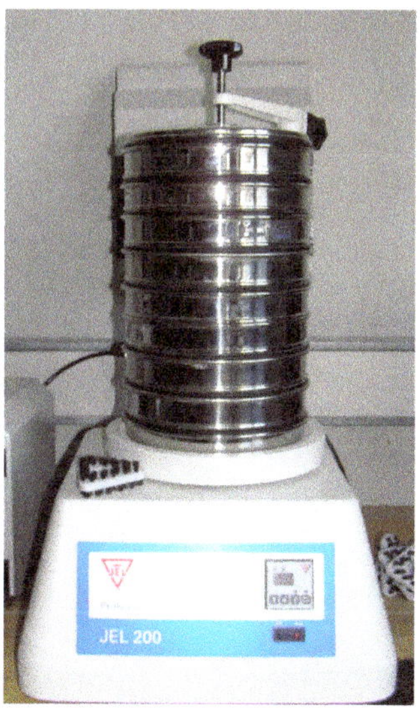

A mechanical shaker used for sieve analysis.

The column is typically placed in a mechanical shaker. The shaker shakes the column, usually for some fixed amount of time. After the shaking is complete the material on each sieve is weighed. The weight of the sample of each sieve is then divided by the total weight to give a percentage retained on each sieve. The size of the average particle on each sieve is then analysed to get a cut-off point or specific size range, which is then captured on a screen.

The results of this test are used to describe the properties of the aggregate and to see if it is appropriate for various civil engineering purposes such as selecting the appropriate aggregate for concrete mixes and asphalt mixes as well as sizing of water production well screens.

The results of this test are provided in graphical form to identify the type of gradation of the aggregate. The complete procedure for this test is outlined in the American Society for Testing and Materials (ASTM) C 136 and the American Association and State Highway and Transportation Officials (AASHTO) T 27

A suitable sieve size for the aggregate underneath the nest of sieves to collect the aggregate that passes through the smallest. The entire nest is then agitated, and the material whose diameter is smaller than the mesh opening pass through the sieves. After the aggregate reaches the pan, the amount of material retained in each sieve is then weighed.

Preparation

In order to perform the test, a sufficient sample of the aggregate must be obtained from the source. To prepare the sample, the aggregate should be mixed thoroughly and be reduced to a suitable size for testing. The total weight of the sample is also required.

Results

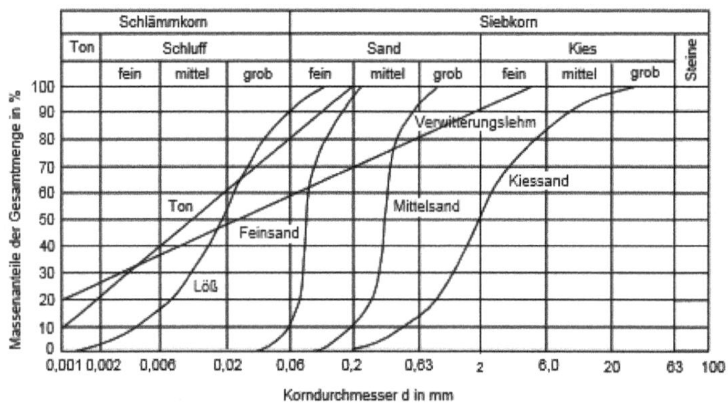

Graphs of cumulative percent passing versus the logarithmic sieve size.

The results are presented in a graph of percent passing versus the sieve size. On the graph the sieve size scale is logarithmic. To find the percent of aggregate passing through each sieve, first find the percent retained in each sieve. To do so, the following equation is used,

$$\%\text{Retained} = \frac{W_{Sieve}}{W_{Total}} \times 100\%$$

where W_{Sieve} is the weight of aggregate in the sieve and W_{Total} is the total weight of the aggregate. The next step is to find the cumulative percent of aggregate retained in each sieve. To do so, add up the total amount of aggregate that is retained in each sieve and the amount in the previous sieves. The cumulative percent passing of the aggregate is found by subtracting the percent retained from 100%.

%Cumulative Passing = 100% - %Cumulative Retained.

The values are then plotted on a graph with cumulative percent passing on the y axis and logarithmic sieve size on the x axis.

There are two versions of the %Passing equations. the .45 power formula is presented on .45 power gradation chart, whereas the more simple %Passing is presented on a semi-log gradation chart. version of the percent passing graph is shown on .45 power chart and by using the .45 passing formula.

.45 power percent passing formula

$$\%\ \text{Passing} = \text{Pi} = \frac{Sieve_{Largest}}{Aggregate_{max-size}}\ \text{x}\ 100\%$$

Where:

$Sieve_{Largest}$ - Largest diameter sieve used in (mm).

$Aggregate_{max\_size}$ - Largest piece of aggregate in the sample in (mm).

Percent Passing Formula

$$\%\text{Passing} = \frac{W_{Below}}{W_{Total}}\ \text{x}\ 100\%$$

Where:

W_{Below} - The total mass of the aggregate within the sieves below the current sieve, not including the current sieve's aggregate.

W_{Total} - The total mass of all of the aggregate in the sample.

Methods

There are different methods for carrying out sieve analyses, depending on the material to be measured.

Throw-action

Here a throwing motion acts on the sample. The vertical throwing motion is overlaid with a slight circular motion which results in distribution of the sample amount over the whole sieving surface. The particles are accelerated in the vertical direction (are thrown upwards). In the air they carry out free rotations and interact with the openings in the mesh of the sieve when they fall back. If the particles are smaller than the openings, they pass through the sieve. If they are larger, they are thrown upwards again. The rotating motion while suspended increases the probability that the particles present a different orientation to the mesh when they fall back again, and thus might eventually pass through the mesh.

Modern sieve shakers work with an electro-magnetic drive which moves a spring-mass system and transfers the resulting oscillation to the sieve stack. Amplitude and sieving time are set digitally and are continuously observed by an integrated control-unit. Therefore, sieving results are reproducible and precise (an important precondition for a significant analysis). Adjustment of parameters like amplitude and sieving time serves to optimize the sieving for different types of material. This method is the most common in the laboratory sector.

Horizontal

In a horizontal sieve shaker the sieve stack moves in horizontal circles in a plane. Horizontal sieve shakers are preferably used for needle-shaped, flat, long or fibrous samples, as their horizontal orientation means that only a few disoriented particles enter the mesh and the sieve is not blocked so quickly. The large sieving area enables the sieving of large amounts of sample, for example as encountered in the particle-size analysis of construction materials and aggregates.

Tapping

Tapping sieving

A horizontal circular motion overlies a vertical motion which is created by a tapping impulse. These motional processes are characteristic of hand sieving and produce a higher degree of sieving for denser particles (e.g. abrasives) than throw-action sieve shakers.

Wet

Most sieve analyses are carried out dry. But there are some applications which can only be carried out by wet sieving. This is the case when the sample which has to be analysed is e.g. a suspension which must not be dried; or when the sample is a very fine powder which tends to agglomerate (mostly < 45 µm) – in a dry sieving process this tendency would lead to a clogging of the sieve meshes and this would make a further sieving process impossible. A wet sieving process is set up like a dry process: the sieve stack is clamped onto the sieve shaker and the sample is placed on the top sieve. Above the top sieve a water-spray nozzle is placed which supports the sieving process additionally to the sieving motion. The rinsing is carried out until the liquid which is discharged through the receiver is clear. Sample residues on the sieves have to be dried and weighed. When it comes to wet sieving it is very important not to change to sample in its volume (no swelling, dissolving or reaction with the liquid).

Air Circular Jet

Air jet sieving machines are ideally suited for very fine powders which tend to agglomerate and cannot be separated by vibrational sieving. The reason for the effectiveness of this sieving method is based on two components: A rotating slotted nozzle inside the sieving chamber and a powerful industrial vacuum cleaner which is connected to the chamber. The vacuum cleaner generates a vacuum inside the sieving chamber and sucks in fresh air through the slotted nozzle. When passing the narrow slit of the nozzle the air stream is accelerated and blown against the sieve mesh, dispersing the particles. Above the mesh, the air jet is distributed over the complete sieve surface and is sucked in with low speed through the sieve mesh. Thus the finer particles are transported through the mesh openings into the vacuum cleaner.

Types of Gradation

A Dense gradation

> A dense gradation refers to a sample that is approximately of equal amounts of various sizes of aggregate. By having a dense gradation, most of the air voids between the material are filled with particles. A dense gradation will result in an even curve on the gradation graph.

Narrow gradation

> Also known as uniform gradation, a narrow gradation is a sample that has aggregate of approximately the same size. The curve on the gradation graph is very steep, and occupies a small range of the aggregate.

Gap gradation

A gap gradation refers to a sample with very little aggregate in the medium size range. This results in only coarse and fine aggregate. The curve is horizontal in the medium size range on the gradation graph.

Open gradation

An open gradation refers an aggregate sample with very little fine aggregate particles. This results in many air voids, because there are no fine particles to fill them. On the gradation graph, it appears as a curve that is horizontal in the small size range.

Rich gradation

A rich gradation refers to a sample of aggregate with a high proportion of particles of small sizes.

Types of Sieves

Woven wire mesh sieves

Woven wire mesh sieves are according to technical requirements of ISO 3310-1. These sieves usually have nominal aperture ranging from 20 micrometers to 3.55 millimeters, with diameters ranging from 100 to 450 millimeters.

Perforated plate sieves

Perforated plate sieves conform to ISO 3310-2 and can have round or square nominal apertures ranging from 1 millimeter to 125 millimeters. The diameters of the sieves range from 200 to 450 millimeters.

American standard sieves

American standard sieves also known as ASTM sieves conform to ASTM E11 standard. The nominal aperture of these sieves range from 20 micrometers to 200 millimeters, however these sieves have only 8 and 12 inch diameter sizes.

Limitations of Sieve Analysis

Sieve analysis has, in general, been used for decades to monitor material quality based on particle size. For coarse material, sizes that range down to #100 mesh (150μm), a sieve analysis and particle size distribution is accurate and consistent.

However, for material that is finer than 100 mesh, dry sieving can be significantly less accurate. This is because the mechanical energy required to make particles pass through an opening and the surface attraction effects between the particles themselves and between particles and the screen increase as the particle size decreases. Wet sieve analysis can be utilized where the material analyzed is not affected by the liquid - except to disperse it. Suspending the particles in a suitable liquid transports fine material through the sieve much more efficiently than shaking the dry material.

Sieve analysis assumes that all particle will be round (spherical) or nearly so and will pass through the square openings when the particle diameter is less than the size of the square opening in the screen. For elongated and flat particles a sieve analysis will not yield reliable mass-based results, as the particle size reported will assume that the particles are spherical, where in fact an elongated particle might pass through the screen end-on, but would be prevented from doing so if it presented itself side-on.

Properties

Gradation affects many properties of an aggregate. It affects bulk density, physical stability and permeability. With careful selection of the gradation, it is possible to achieve high bulk density, high physical stability, and low permeability. This is important because in pavement design, a workable, stable mix with resistance to water is important. With an open gradation, the bulk density is relatively low, due to the lack of fine particles, the physical stability is moderate, and the permeability is quite high. With a rich gradation, the bulk density will also be low, the physical stability is low, and the permeability is also low. The gradation can be affected to achieve the desired properties for the particular engineering application.

Engineering Applications

Gradation is usually specified for each engineering application it is used for. For example, foundations might only call for coarse aggregates, and therefore an open gradation is needed.

Sieving with Digital Image Processing

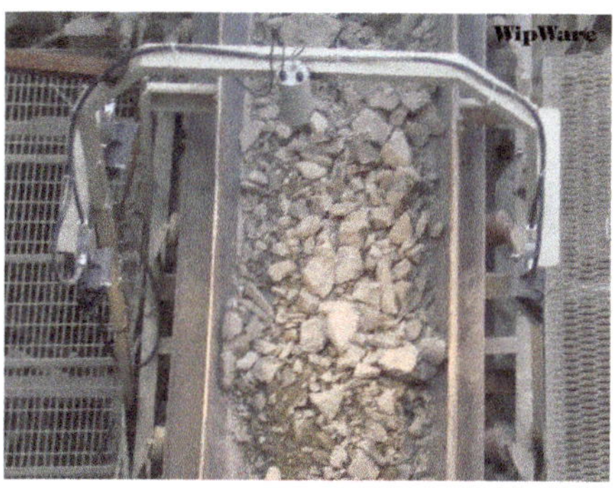

Digital image processing to determine PSD has now become the quickest and easiest method for analyzing rock masses. The algorithms used to analyze digital images and produce PSD has been researched and developed at the University of Arizona since 1997, and now in software offered by multiple companies around the world.

The Method for Digital Image Processing is The Following

Image acquisition: Using a standard digital camera, a user takes a picture of the rocks for analysis. Loading image into the program: Typically user friendly, either through File or drag and drop Processing image with software's proprietary algorithm: The software analyzes individual pixel's gradient and compares to each of the neighboring pixels to determine boundaries between the rocks. Editing: User edits the boundaries to better accuracy Results: PSD is produced by the program. The results page is typically customizable for reporting units and graph style.

Real Time Data Reporting of Image Processing

Also available is real time data reporting from any desired location within the comminution circuit. Continuous conveyor belt fragmentation measurements in the comminution plant can provide an instantaneous tool for process control. When planning the integration of fragmentation measurement instrumentation, position your camera monitoring station in a location where you can effect a process control change based upon the particle size information you receive. For example, the primary crusher product is a good location to measure, since you are able to control the size that the crusher is producing with slight adjustments to the crusher ensuring that the downstream comminution circuit is receiving a consistent product to help reduce variability in the process. In addition to the continuous particle size analysis, the database of belt material size can be used for simulations and for further process analysis to help increase process efficiency and throughput.

Grading Characteristics

A grading curve is a useful aid to soil description. The geometric properties of a grading curve are called grading characteristics.

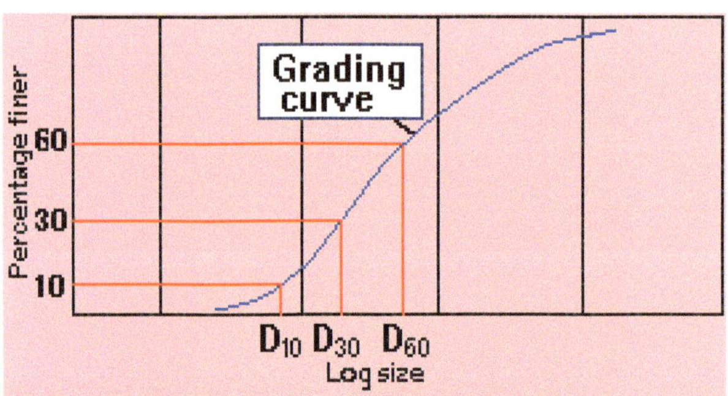

To obtain the grading characteristics, three points are located first on the grading curve.

D_{60} = size at 60% finer by weight

D_{30} = size at 30% finer by weight

D_{10} = size at 10% finer by weight

The grading characteristics are then determined as follows:

1. Effective size = D_{10}

2. Uniformity coefficient, $C_u = \frac{D_{60}}{D_{10}}$

3. Curvature coefficient, $C_c = \frac{(D_{30})^2}{D_{60} D_{10}}$

Both C_u and C_c will be 1 for a single-sized soil.

$C_u > 5$ indicates a well-graded soil, i.e. a soil which has a distribution of particles over a wide size range.

C_c between 1 and 3 also indicates a well-graded soil.

$C_u < 3$ indicates a uniform soil, i.e. a soil which has a very narrow particle size range.

Consistency of Soils

The consistency of a fine-grained soil refers to its firmness, and it varies with the water content of the soil.

A gradual increase in water content causes the soil to change from *solid* to *semi-solid* to *plastic* to *liquid* states. The water contents at which the consistency changes from one state to the other are called consistency limits (or Atterberg limits).

The three limits are known as the shrinkage limit (W_S), plastic limit (W_P), and liquid limit (W_L) as shown. The values of these limits can be obtained from laboratory tests.

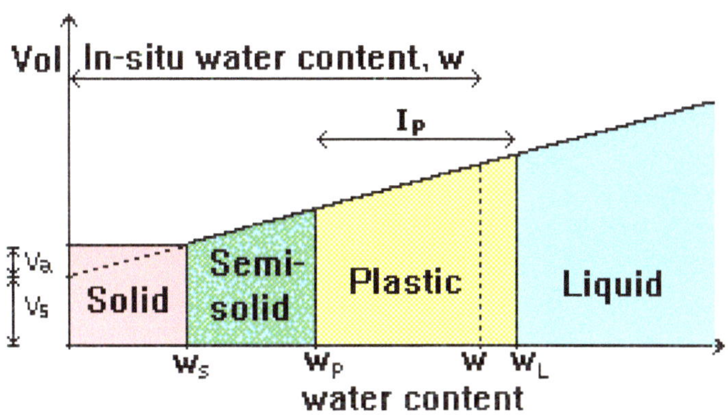

Two of these are utilised in the classification of fine soils:

Liquid limit (W_L) - change of consistency from plastic to liquid state

Plastic limit (W_P) - change of consistency from brittle/crumbly to plastic state

The difference between the liquid limit and the plastic limit is known as the plasticity index (I_p), and it is in this range of water content that the soil has a plastic consistency. The consistency of most soils in the field will be plastic or semi-solid.

Atterberg Limits

The Atterberg limits are a basic measure of the critical water contents of a fine-grained soil: its shrinkage limit, plastic limit, and liquid limit. As a dry, clayey soil takes on increasing amounts of water, it undergoes distinct changes in behavior and consistency. Depending on the water content of the soil, it may appear in four states: solid, semi-solid, plastic and liquid. In each state, the consistency and behavior of a soil is different and consequently so are its engineering properties. Thus, the boundary between each state can be defined based on a change in the soil's behavior. The Atterberg limits can be used to distinguish between silt and clay, and it can distinguish between different types of silts and clays. These limits were created by Albert Atterberg, a Swedish agriculturist. They were later refined by Arthur Casagrande. These distinctions in soil are used in assessing the soils that are to have structures built on. Soils when wet retain water and some expand in volume. The amount of expansion is related to the ability of the soil to take in water and its structural make-up (the type of atoms present). These tests are mainly used on clayey or silty soils since these are the soils that expand and shrink due to moisture content. Clays and silts react with the water and thus change sizes and have varying shear strengths. Thus these tests are used widely in the preliminary stages of designing any structure to ensure that the soil will have the correct amount of shear strength and not too much change in volume as it expands and shrinks with different moisture contents.

As a hard, rigid solid in the dry state, soil becomes a crumbly (friable) semisolid when a certain moisture content, termed the shrinkage limit, is reached. If it is an expansive soil, this soil will also begin to swell in volume as this moisture content is exceeded. Increasing the water content beyond the soil's plastic limit will transform it into a malleable, plastic mass, which causes additional swelling. The soil will remain in this plastic state until its liquid limit is exceeded, which causes it to transform into a viscous liquid that flows when jarred.

Laboratory Tests

Shrinkage Limit

The shrinkage limit (SL) is the water content where further loss of moisture will not result in any more volume reduction. The test to determine the shrinkage limit is ASTM International D4943. The shrinkage limit is much less commonly used than the liquid and plastic limits.

Plastic Limit

The plastic limit (PL) is determined by rolling out a thread of the fine portion of a soil on a flat, non-porous surface. The procedure is defined in ASTM Standard D 4318. If the soil is at a moisture content where its behavior is plastic, this thread will retain its shape down to a very narrow diameter. The sample can then be remolded and the test repeated. As the moisture content falls due to evaporation, the thread will begin to break apart at larger diameters. The plastic limit is defined as the moisture content where the thread breaks apart at a diameter of 3.2 mm (about 1/8 inch). A soil is considered non-plastic if a thread cannot be rolled out down to 3.2 mm at any moisture possible.

Liquid Limit

The liquid limit (LL) is conceptually defined as the water content at which the behavior of a clayey soil changes from plastic to liquid. However, the transition from plastic to liquid behavior is gradual over a range of water contents, and the shear strength of the soil is not actually zero at the liquid limit. The precise definition of the liquid limit is based on standard test procedures described below.

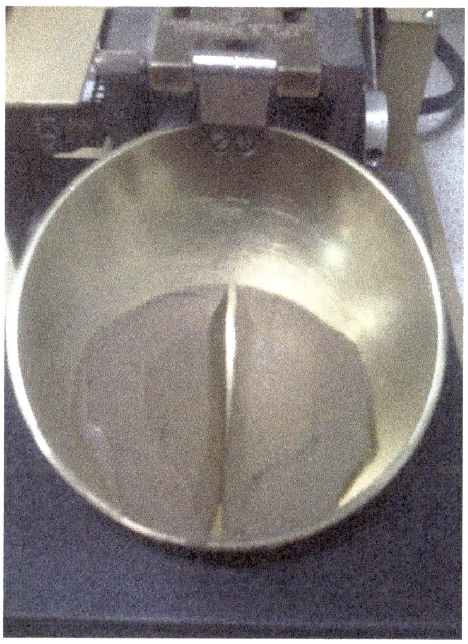

Casagrande cup in action

The original liquid limit test of Atterberg's involved mixing a pat of clay in a round-bottomed porcelain bowl of 10–12 cm diameter. A groove was cut through the pat of clay with a spatula, and the bowl was then struck many times against the palm of one hand. Casagrande subsequently standardized the apparatus and the procedures to make the measurement more repeatable. Soil is placed into the metal cup portion of the device and a groove is made down its center with a standardized tool of 13.5 millimetres (0.53 in) width. The cup is repeatedly dropped 10 mm onto a hard rubber base at a rate of 120 blows per minute, during which the groove closes up gradually as a result of the impact. The number of blows for the groove to close is recorded. The moisture content at which it takes 25 drops of the cup to cause the groove to close over a distance of 13.5

millimetres (0.53 in) is defined as the liquid limit. The test is normally run at several moisture contents, and the moisture content which requires 25 blows to close the groove is interpolated from the test results. The Liquid Limit test is defined by ASTM standard test method D 4318. The test method also allows running the test at one moisture content where 20 to 30 blows are required to close the groove; then a correction factor is applied to obtain the liquid limit from the moisture content.

Another method for measuring the liquid limit is the fall cone test, also called the cone penetrometer test. It is based on the measurement of penetration into the soil of a standardized cone of specific mass. Although the Casagrande test is widely used across North America, the fall cone test is much more prevalent in Europe due to being less dependent on the operator in determining the Liquid Limit.

Derived Limits

The values of these limits are used in a number of ways. There is also a close relationship between the limits and properties of a soil such as compressibility, permeability, and strength. This is thought to be very useful because as limit determination is relatively simple, it is more difficult to determine these other properties. Thus the Atterberg limits are not only used to identify the soil's classification, but it allows for the use of empirical correlations for some other engineering properties.

Plasticity Index

The plasticity index (PI) is a measure of the plasticity of a soil. The plasticity index is the size of the range of water contents where the soil exhibits plastic properties. The PI is the difference between the liquid limit and the plastic limit (PI = LL-PL). Soils with a high PI tend to be clay, those with a lower PI tend to be silt, and those with a PI of 0 (non-plastic) tend to have little or no silt or clay.

Soil descriptions based on PI

- (0-3)- Nonplastic
- (3-15) - Slightly plastic
- (15-30) - Medium plastic
- >30 - Highly plastic

Liquidity Index

The liquidity index (LI) is used for scaling the natural water content of a soil sample to the limits. It can be calculated as a ratio of difference between natural water content, plastic limit, and liquid limit: LI=(W-PL)/(LL-PL) where W is the natural water content.

Consistency Index

The consistency index (CI) indicates the consistency (firmness) of a soil. It is calculated as CI = (LL-W)/(LL-PL), where W is the existing water content. Soil at the liquid limit will have a consistency index of 0, while soil at the plastic limit will have a consistency index of 1.

Activity

The activity (A) of a soil is the PI divided by the percent of clay-sized particles (less than 2 μm) present. Different types of clays have different specific surface areas which controls how much wetting is required to move a soil from one phase to another such as across the liquid limit or the plastic limit. From the activity, one can predict the dominant clay type present in a soil sample. High activity signifies large volume change when wetted and large shrinkage when dried. Soils with high activity are very reactive chemically. Normally the activity of clay is between 0.75 and 1.25, and in this range clay is called normal. It is assumed that the plasticity index is approximately equal to the clay fraction (A = 1). When A is less than 0.75, it is considered inactive. When it is greater than 1.25, it is considered active.

References

- Krumbein, W. C.; Aberdeen, Esther (April 1937). "The Sediments of Barataria Bay". Journal of Sedimentary Petrology. 7 (1). doi:10.1306/D4268F8B-2B26-11D7-8648000102C1865D. Retrieved 11 May 2014
- "Soil classification system of England and Wales". Cranfield University, National Soil Resources Institute. Retrieved 2011-12-22
- Folk, Robert L.; Ward, William C. (1957). "Brazos River bar: a study in the significance of grain-size parameters" (PDF). Journal of Sedimentary Petrology. 27 (1): 3–26. doi:10.1306/74d70646-2b21-11d7-8648000102c1865d. Retrieved 11 May 2014
- Texas Department of Transportation (January 2016). "Test Procedure for Sieve Analysis of Fine and Coarse Aggregates" (PDF). Texas DOT. Retrieved 2016-12-24
- Seed, H.B. (1967). "Fundamental Aspects of the Atterberg Limits". Journal of Soil Mechanics and Foundations Div., 92(SM4)

2

Soil Properties: An Overview

Soil phases relate to the composition of soil which is classified into solids, liquids and gases. Soil analysis calculates porosity, void ratio, air ratio and the degree of saturation with regard to the different soil phases. Soil fertility is the ability of soil to provide a satisfactory habitat and productive yield to a plant. It depends on soil's ability to provide nutrients and water, and also on the absence of toxic substances. This chapter has been carefully written to provide an easy understanding of the varied facets of soil mechanics.

Soil Phases

Soil is not a coherent solid material like steel and concrete, but is a particulate material. Soils, as they exist in nature, consist of solid particles (mineral grains, rock fragments) with water and air in the voids between the particles. The water and air contents are readily changed by changes in ambient conditions and location.

As the relative proportions of the three phases vary in any soil deposit, it is useful to consider a soil model which will represent these phases distinctly and properly quantify the amount of each phase. A schematic diagram of the three-phase system is shown in terms of weight and volume symbols respectively for soil solids, water, and air. The weight of air can be neglected.

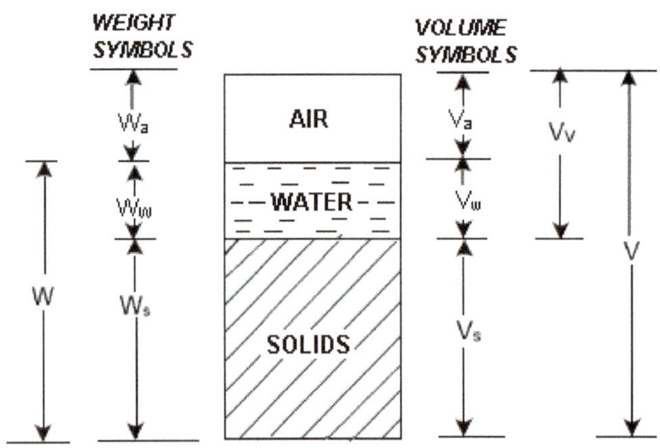

The soil model is given dimensional values for the solid, water and air components.

Total volume, $V = V_s + V_w + V_v$

Three-phase System

Soils can be partially saturated (with both air and water present), or be fully saturated (no air content) or be perfectly dry (no water content).

In a saturated soil or a dry soil, the three-phase system thus reduces to two phases only, as shown.

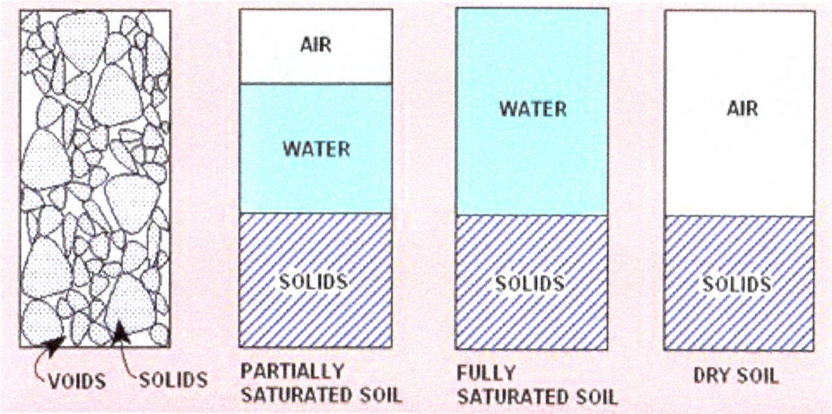

For the purpose of engineering analysis and design, it is necessary to express relations between the weights and the volumes of the three phases.

The various relations can be grouped into:

- Volume relations
- Weight relations
- Inter-relations

Void Ratio

Void ratio, in materials science, is a quantity related to porosity and defined as the ratio:

$$e = \frac{V_V}{V_S} = \frac{V_V}{V_T - V_V} = \frac{\phi}{1-\phi}$$

and

$$\phi = \frac{V_V}{V_T} = \frac{V_V}{V_S + V_V} = \frac{e}{1+e}$$

where e is void ratio, ϕ is porosity, V_V is the volume of void-space (such as fluids), V_S is the volume of solids, and V_T is the total or bulk volume. This figure is relevant in composites, in mining (particular with regard to the properties of tailings), and in soil science. In geotechnical engineering, it is considered as one of the state variables of soils and represented by the symbol e.

Note that in geotechnical engineering, the symbol ϕ usually represents the angle of shearing resistance, a shear strength (soil) parameter. Because of this, the equation is usually rewritten using n for porosity:

$$e = \frac{V_V}{V_S} = \frac{V_V}{V_T - V_V} = \frac{n}{1-n}$$

and

$$n = \frac{V_V}{V_T} = \frac{V_V}{V_S + V_V} = \frac{e}{1+e}$$

where e is void ratio, n is porosity, VV is the volume of void-space (air and water), VS is the volume of solids, and VT is the total or bulk volume.

Engineering Applications

- Volume change tendency control. If void ratio is high (loose soils) voids in a soil skeleton tend to minimize under loading - adjacent particles contract. The opposite situation, i.e. when void ratio is relatively small (dense soils), indicates that the volume of the soil is vulnerable to increase under loading - particles dilate.
- Fluid conductivity control (ability of water movement through the soil). Loose soils show high conductivity, while dense soils are not so permeable.
- Particles movement. In a loose soil particles can move quite easily, whereas in a dense one finer particles cannot pass through the voids, which leads to clogging.

Pore Space in Soil

The pore space of soil contains the liquid and gas phases of soil, i.e., everything but the solid phase that contains mainly minerals of varying sizes as well as organic compounds.

In order to understand porosity better a series of equations have been used to express the quantitative interactions between the three phases of soil.

Macropores or fractures play a major role in infiltration rates in many soils as well as preferential flow patterns, hydraulic conductivity and evapotranspiration. Cracks are also very influential in gas exchange, influencing respiration within soils. Modeling cracks therefore helps understand how these processes work and what the effects of changes in soil cracking such as compaction, can have on these processes.

Background

Bulk Density

$$\rho = \frac{M_s}{V_t}$$

The bulk density of soil depends greatly on the mineral make up of soil and the degree of compaction. The density of quartz is around 2.65 g/cm^3 but the bulk density of a soil may be less than half that density.

Most soils have a bulk density between 1.0 and 1.6 g/cm^3 but organic soil and some friable clay may have a bulk density well below 1 g/cm^3

Core samples are taken by driving a metal core into the earth at the desired depth and soil horizon. The samples are then oven dried and weighed.

Bulk density = (mass of oven dry soil)/volume

The bulk density of soil is inversely related to the porosity of the same soil. The more pore space in a soil the lower the value for bulk density.

Porosity

$$f = \frac{V_f}{V_t} \text{ or } f = \frac{V_a + V_w}{V_s + V_a + V_w}$$

Porosity is a measure of the total pore space in the soil. This is measured as a volume or percent. The amount of porosity in a soil depends on the minerals that make up the soil and the amount of sorting that occurs within the soil structure. For example, a sandy soil will have larger porosity than silty sand, because the silt will fill in the gaps between the sand particles.

Pore Space Relations

Hydraulic Conductivity

Hydraulic conductivity (K) is a property of soil that describes the ease with which water can move through pore spaces. It depends on the permeability of the material (pores, compaction) and on the degree of saturation. Saturated hydraulic conductivity, K_{sat}, describes water movement through saturated media. Where hydraulic conductivity has the capability to be measured at any state. It can be estimated by numerous kinds of equipment. To calculate hydraulic conductivity, Darcy's law is used. The manipulation of the law depends on the Soil saturation and instrument used.

Infiltration

Infiltration is the process by which water on the ground surface enters the soil. The water enters the soil through the pores by the forces of gravity and capillary action. The largest cracks and pores offer a great reservoir for the initial flush of water. This allows a rapid infiltration. The smaller pores take longer to fill and rely on capillary forces as well as gravity. The smaller pores have a slower infiltration as the soil becomes more saturated.

Pore Types

A pore is not simply a void in the solid structure of soil. The various pore size categories have different characteristics and contribute different attributes to soils depending on the number and frequency of each type. A widely used classification of pore size is that of Brewer (1964):

Macropore

The pores that are too large to have any significant capillary force. Unless impeded, water will drain from these pores, and they are generally air-filled at field capacity. Macropores can be caused by cracking, division of peds and aggregates, as well as plant roots, and zoological exploration. Size >75 μm.

Mesopore

The largest pores filled with water at field capacity. Also known as storage pores because of the ability to store water useful to plants. They do not have capillary forces too great so that the water does not become limiting to the plants. The properties of mesopores are highly studied by soil scientists because of their impact on agriculture and irrigation. Size 30 μm–75 μm.

Micropore

These are "pores that are sufficiently small that water within these pores is considered immobile, but available for plant extraction." Because there is little movement of water in these pores, solute movement is mainly by the process of diffusion. Size 5-30 μm.

Ultramicropore

These pores are suitable for habitation by microorganisms. Their distribution is determined by soil texture and soil organic matter, and they are not greatly affected by compaction Size 0.1-30 μm.

Cryptopore

Pores that are too small to be penetrated by most microorganisms. Organic matter in these pores is therefore protected from microbial decomposition. They are filled with water unless the soil is very dry, but little of this water is available to plants, and water movement is very slow. Size <0.1 μm.

Modelling Methods

Basic crack modeling has been undertaken for many years by simple observations and measurements of crack size, distribution, continuity and depth. These observations have either been surface observation or done on profiles in pits. Hand tracing and measurement of crack patterns on paper was one method used prior to advances in modern technology. Another field method was with the use of string and a semicircle of wire. The semi circle was moved along alternating sides of a string line. The cracks within the semicircle were measured for width, length and depth using a ruler. The crack distribution was calculated using the principle of Buffon's needle.

Disc Permeameter

This method relies on the fact that crack sizes have a range of different water potentials. At zero water potential at the soil surface an estimate of saturated hydraulic conductivity is produced, with all pores filled with water. As the potential is decreased progressively larger cracks drain. By measuring at the hydraulic conductivity at a range of negative potentials, the pore size distribution can be determined. While this is not a physical model of the cracks, it does give an indication to the sizes of pores within the soil.

Horgan and Young Model

Horgan and Young (2000) produced a computer model to create a two-dimensional prediction of surface crack formation. It used the fact that once cracks come within a certain distance of one another they tend to be attracted to each other. Cracks also tend to turn within a particular range of angles and at some stage a surface aggregate gets to a size that no more cracking will occur. These are often characteristic of a soil and can therefore be measured in the field and used in the model. However it was not able to predict the points at which cracking starts and although random in the formation of crack pattern, in many ways, cracking of soil is often not random, but follows lines of weaknesses.

Araldite-impregnation Imaging

A large core sample is collected. This is then impregnated with araldite and a fluorescent resin. The core is then cut back using a grinding implement, very gradually

(~1 mm per time), and at every interval the surface of the core sample is digitally imaged. The images are then loaded into a computer where they can be analysed. Depth, continuity, surface area and a number of other measurements can then be made on the cracks within the soil.

Electrical Resistivity Imaging

Using the infinite resistivity of air, the air spaces within a soil can be mapped. A specially designed resistivity meter had improved the meter-soil contact and therefore the area of the reading. This technology can be used to produce images that can be analysed for a range of cracking properties.

Weight Relations

Density is a measure of the quantity of mass in a unit volume of material. Unit weight is a measure of the weight of a unit volume of material. Both can be used interchangeably. The units of density are ton/m³, kg/m³ or g/cm³. The following are the basic weight relations:

1. The ratio of the mass of water present to the mass of solid particles is called the water content (w), or sometimes the moisture content.

$$w = \frac{W_w}{W_S}$$

Its value is 0% for dry soil and its magnitude can exceed 100%.

2. The mass of solid particles is usually expressed in terms of their particle unit weight (y_s) or specific gravity (Gs) of the soil grain solids.

$$\gamma_s = \frac{W_s}{V_s} = G_s \cdot \gamma_W$$

where γ_W = Unit weight of water.

For most inorganic soils, the value of Gs lies between 2.60 and 2.80. The presence of organic material reduces the value of Gs.

3. Dry unit weight (γ_d) is a measure of the amount of solid particles per unit volume

$$\gamma_d = \frac{W_s}{V}$$

4. Bulk unit weight $(\gamma_s \; or \; \gamma)$ is a measure of the amount of solid particles plus water per unit volume.

$$\gamma_t = \gamma = \frac{(W_s + W_W)}{(V_s + V_V)}$$

5. Saturated unit weight (γ_{502}) is equal to the bulk density when the total voids is filled up with water.

6. Buoyant unit weight (γ') or submerged unit weight is the effective mass per unit volume when the soil is submerged below standing water or below the ground water table.

$$\gamma' = \gamma_{502} - \gamma_W$$

It is important to quantify the state of a soil immediately after receiving in the laboratory and prior to commencing other tests. The water content and unit weight are particularly important, since they may change during transportation and storage.

Some physical state properties are calculated following the practical measurement of others. For example, dry unit weight can be determined from bulk unit weight and water content. The following are some inter-relations:

1. $w = \frac{W_W}{W_s} = \frac{\gamma_W \cdot V_W}{G_s \cdot \gamma_W \cdot V_s} = \frac{V_W}{G_s \cdot V_s} = \frac{S.V_v}{G_s.V_s} = \frac{S.e}{G_s}$

2. $\gamma = \frac{(G_s + S.e.)\gamma_w}{1+e}$

3. $\gamma = \frac{(1+w).G_s.\gamma_w}{1+e}$

4. $\gamma_d = \frac{G_s.\gamma_W}{1+e}$

5. $\gamma_d = \frac{\gamma}{1+w}$

6. $\gamma' = \frac{[(G_s - 1) + (S-1)e]xy_w}{1+e}$

7. $\gamma' = \frac{(G_s - 1).y_w}{1+e}$

Soil Fertility

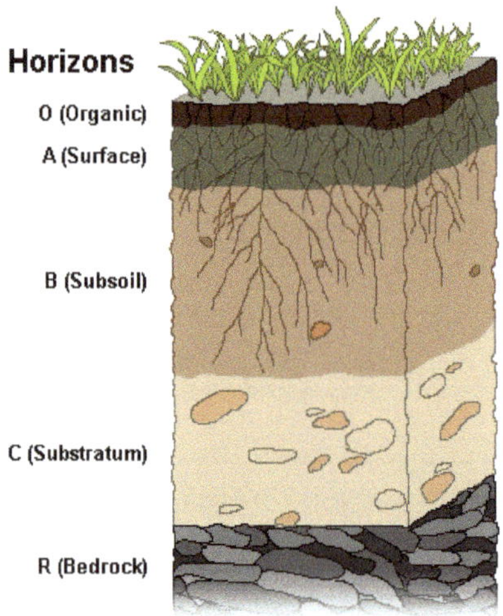

Soil scientists use the capital letters O, A, B, C, and E to identify the master horizons, and lowercase letters for distinctions of these horizons. Most soils have three major horizons—the surface horizon (A), the subsoil (B), and the substratum (C). Some soils have an organic horizon (O) on the surface, but this horizon can also be buried. The master horizon, E, is used for subsurface horizons that have a significant loss of minerals (eluviation). Hard bedrock, which is not soil, uses the letter R.

Soil fertility refers to the ability of a soil to sustain agricultural plant growth, i.e. to provide plant habitat and result in sustained and consistent yields of high quality. A fertile soil has the following properties:

- The ability to supply essential plant nutrients and soil water in adequate amounts and proportions for plant growth and reproduction; and
- The absence of toxic substances which may inhibit plant growth.

The following properties contribute to soil fertility in most situations:

- Sufficient soil depth for adequate root growth and water retention;
- Good internal drainage, allowing sufficient aeration for optimal root growth (although some plants, such as rice, tolerate waterlogging);
- Topsoil with sufficient soil organic matter for healthy soil structure and soil moisture retention;
- Soil pH in the range 5.5 to 7.0 (suitable for most plants but some prefer or tolerate more acid or alkaline conditions);
- Adequate concentrations of essential plant nutrients in plant-available forms;
- Presence of a range of microorganisms that support plant growth.

In lands used for agriculture and other human activities, maintenance of soil fertility typically requires the use of soil conservation practices. This is because soil erosion and other forms of soil degradation generally result in a decline in quality with respect to one or more of the aspects indicated above.

Soil Fertilization

Bioavailable phosphorus is the element in soil that is most often lacking. Nitrogen and potassium are also needed in substantial amounts. For this reason these three elements are always identified on a commercial fertilizer analysis. For example, a 10-10-15 fertilizer has 10 percent nitrogen, 10 percent (P_2O_5) available phosphorus and 15 percent (K_2O) water-soluble potassium. Sulfur is the fourth element that may be identified in a commercial analysis—e.g. 21-0-0-24 which would contain 21% nitrogen and 24% sulfate.

Inorganic fertilizers are generally less expensive and have higher concentrations of nutrients than organic fertilizers. Also, since nitrogen, phosphorus and potassium generally must be in the inorganic forms to be taken up by plants, inorganic fertilizers are generally immediately bioavailable to plants without modification. However, some have criticized the use of inorganic fertilizers, claiming that the water-soluble nitrogen doesn't provide for the long-term needs of the plant and creates water pollution. Slow-release fertilizers may reduce leaching loss of nutrients and may make the nutrients that they provide available over a longer period of time.

Soil fertility is a complex process that involves the constant cycling of nutrients between organic and inorganic forms. As plant material and animal wastes are decomposed by micro-organisms, they release inorganic nutrients to the soil solution, a process referred to as mineralization. Those nutrients may then undergo further transformations which may be aided or enabled by soil micro-organisms. Like plants, many micro-organisms require or preferentially use inorganic forms of nitrogen, phosphorus or potassium and will compete with plants for these nutrients, tying up the nutrients in microbial biomass, a process often called immobilization. The balance between immobilization and mineralization processes depends on the balance and availability of major nutrients and organic carbon to soil microorganisms. Natural processes such as lightning strikes may fix atmospheric nitrogen by converting it to (NO_2). Denitrification may occur under anaerobic conditions (flooding) in the presence of denitrifying bacteria. Nutrient cations, including potassium and many micronutrients, are held in relatively strong bonds with the negatively charged portions of the soil in a process known as cation exchange.

In 2008 the cost of phosphorus as fertilizer more than doubled, while the price of rock phosphate as base commodity rose eight-fold. Recently the term peak phosphorus has been coined, due to the limited occurrence of rock phosphate in the world.

Light and CO_2 limitations

Photosynthesis is the process whereby plants use light energy to drive chemical reactions which convert CO_2 into sugars. As such, all plants require access to both light and carbon dioxide to produce energy, grow and reproduce.

While typically limited by nitrogen, phosphorus and potassium, low levels of carbon dioxide can also act as a limiting factor on plant growth. Peer-reviewed and published scientific studies have shown that increasing CO_2 is highly effective at promoting plant growth up to levels over 300 ppm. Further increases in CO_2 can, to a very small degree, continue to increase net photosynthetic output.

Since higher levels of CO_2 have only a minimal impact on photosynthetic output at present levels (presently around 400 ppm and increasing), we should not consider plant growth to be limited by carbon dioxide. Other biochemical limitations, such as soil organic content, nitrogen in the soil, phosphorus and potassium, are far more often in short supply. As such, neither commercial nor scientific communities look to air fertilization as an effective or economic method of increasing production in agriculture or natural ecosystems. Furthermore, since microbial decomposition occurs faster under warmer temperatures, higher levels of CO_2 (which is one of the causes of unusually fast climate change) should be expected to increase the rate at which nutrients are leached out of soils and may have a negative impact on soil fertility.

Soil Depletion

Soil depletion occurs when the components which contribute to fertility are removed and not replaced, and the conditions which support soil's fertility are not maintained. This leads to poor crop yields. In agriculture, depletion can be due to excessively intense cultivation and inadequate soil management.

Soil fertility can be severely challenged when land use changes rapidly. For example, in Colonial New England, colonists made a number of decisions that depleted the soils, including: allowing herd animals to wander freely, not replenishing soils with manure, and a sequence of events that led to erosion. William Cronon wrote that "...the long-term effect was to put those soils in jeopardy. The removal of the forest, the increase in destructive floods, the soil compaction and close-cropping wrought by grazing animals, plowing--all served to increase erosion."

Karl Marx wrote of the role of capitalism in soil depletion. In *Capital, Volume I*, he wrote:

> ...all progress in capitalistic agriculture is a progress in the art, not only of robbing the labourer, but of robbing the soil; all progress in increasing the fertility of the soil for a given time, is a progress towards ruining the lasting sources of that fertility. The more a country starts its development on the foundation of

> modern industry, like the United States, for example, the more rapid is this process of destruction. Capitalist production, therefore, develops technology, and the combining together of various processes into a social whole, only by sapping the original sources of all wealth — the soil and the labourer.

One of the most widespread occurrences of soil depletion as of 2008 is in tropical zones where nutrient content of soils is low. The combined effects of growing population densities, large-scale industrial logging, slash-and-burn agriculture and ranching, and other factors, have in some places depleted soils through rapid and almost total nutrient removal.

Topsoil depletion occurs when the nutrient-rich organic topsoil, which takes hundreds to thousands of years to build up under natural conditions, is eroded or depleted of its original organic material. Historically, many past civilizations' collapses can be attributed to the depletion of the topsoil. Since the beginning of agricultural production in the Great Plains of North America in the 1880s, about one-half of its topsoil has disappeared.

Depletion may occur through a variety of other effects, including overtillage (which damages soil structure), underuse of nutrient inputs which leads to mining of the soil nutrient bank, and salinization of soil.

Irrigation Water Effects

The quality of irrigation water is very important to maintain soil fertility and tilth, and for using more soil depth by the plants. When soil is irrigated with high alkaline water, unwanted sodium salts build up in the soil which would make soil draining capacity very poor. So plant roots can not penetrate deep into the soil for optimum growth in Alkali soils. When soil is irrigated with low pH / acidic water, the useful salts (Ca, Mg, K, P, S, etc.) are removed by draining water from the acidic soil and in addition unwanted aluminium and manganese salts to the plants are dissolved from the soil impeding plant growth. When soil is irrigated with high salinity water or sufficient water is not draining out from the irrigated soil, the soil would convert into saline soil or lose its fertility. Saline water enhance the turgor pressure or osmotic pressure requirement which impedes the off take of water and nutrients by the plant roots.

Top soil loss takes place in alkali soils due to erosion by rain water surface flows or drainage as they form colloids (fine mud) in contact with water. Plants absorb water-soluble inorganic salts only from the soil for their growth. Soil as such does not lose fertility just by growing crops but it lose its fertility due to accumulation of unwanted and depletion of wanted inorganic salts from the soil by improper irrigation and acid rain water (quantity and quality of water). The fertility of many soils which are not suitable for plant growth can be enhanced many times gradually by providing adequate irrigation water of suitable quality and good drainage from the soil.

Global Distribution

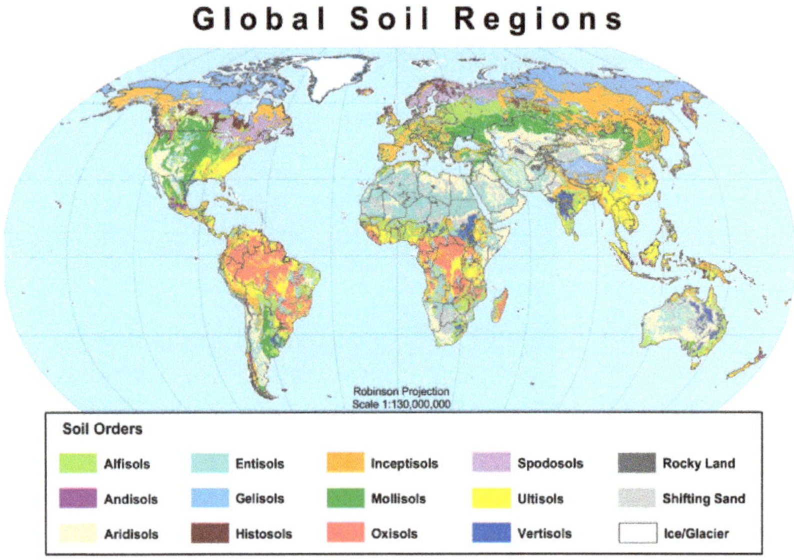

Global distribution of soil types of the USDA soil taxonomy system. Mollisols, shown here in dark green, are a good (though not the only) indicator of high soil fertility. They coincide to a large extent with the world's major grain producing areas like the North American Prairie States, the Pampa and Gran Chaco of South America and the Ukraine-to-Central Asia Black Earth belt.

Fertilizer

A large, modern fertilizer spreader

A fertilizer or fertiliser is any material of natural or synthetic origin (other than liming materials) that is applied to soils or to plant tissues (usually leaves) to supply one or more plant nutrients essential to the growth of plants.

A Lite-Trac Agri-Spread lime and fertilizer spreader at an agricultural show

Mechanism

Six tomato plants grown with and without nitrate fertilizer on nutrient-poor sand/clay soil. One of the plants in the nutrient-poor soil has died.

Fertilizers enhance the growth of plants. This goal is met in two ways, the traditional one being additives that provide nutrients. The second mode by which some fertilizers act is to enhance the effectiveness of the soil by modifying its water retention and aeration. Like many on fertilizers, emphasises the nutritional aspect. Fertilizers typically provide, in varying proportions:

- three main macronutrients:
    - Nitrogen (N): leaf growth;
    - Phosphorus (P): Development of roots, flowers, seeds, fruit;
    - Potassium (K): Strong stem growth, movement of water in plants, promotion of flowering and fruiting;
- three secondary macronutrients: calcium (Ca), magnesium (Mg), and sulphur (S);
- micronutrients: copper (Cu), iron (Fe), manganese (Mn), molybdenum (Mo), zinc (Zn), boron (B), and of occasional significance there are silicon (Si), cobalt (Co), and vanadium (V) plus rare mineral catalysts.

The nutrients required for healthy plant life are classified according to the elements, but the elements are not used as fertilizers. Instead compounds containing these elements are the basis of fertilizers. The macronutrients are consumed in larger quantities and are present in plant tissue in quantities from 0.15% to 6.0% on a dry matter (DM)

(0% moisture) basis. Plants are made up of four main elements: hydrogen, oxygen, carbon, and nitrogen. Carbon, hydrogen and oxygen are widely available as water and carbon dioxide. Although nitrogen makes up most of the atmosphere, it is in a form that is unavailable to plants. Nitrogen is the most important fertilizer since nitrogen is present in proteins, DNA and other components (e.g., chlorophyll). To be nutritious to plants, nitrogen must be made available in a "fixed" form. Only some bacteria and their host plants (notably legumes) can fix atmospheric nitrogen (N_2) by converting it to ammonia. Phosphate is required for the production of DNA and ATP, the main energy carrier in cells, as well as certain lipids.

Micronutrients are consumed in smaller quantities and are present in plant tissue on the order of parts-per-million (ppm), ranging from 0.15 to 400 ppm DM, or less than 0.04% DM. These elements are often present at the active sites of enzymes that carry out the plant's metabolism. Because these elements enable catalysts (enzymes) their impact far exceeds their weight percentage.

Classification

Fertilizers are classified in several ways. They are classified according to whether they provide a single nutrient (say, N, P, or K), in which case they are classified as "straight fertilizers." "Multinutrient fertilizers" (or "complex fertilizers") provide two or more nutrients, for example N and P. Fertilizers are also sometimes classified as inorganic versus organic. Inorganic fertilizers exclude carbon-containing materials except ureas. Organic fertilizers are usually (recycled) plant- or animal-derived matter. Inorganic are sometimes called synthetic fertilizers since various chemical treatments are required for their manufacture.

Single Nutrient ("Straight") Fertilizers

fertilizer is ammonia or its solutions. Ammonium nitrate (NH_4NO_3) is also widely used. Urea is another popular source of nitrogen, having the advantage that it is solid and non-explosive, unlike ammonia and ammonium nitrate, respectively. A few percent of the nitrogen fertilizer market (4% in 2007) has been met by calcium ammonium nitrate ($Ca(NO_3)_2 \cdot NH_4NO_3 \cdot 10H_2O$).

The main straight phosphate fertilizers are the superphosphates. "Single superphosphate" (SSP) consists of 14–18% P_2O_5, again in the form of $Ca(H_2PO_4)_2$, but also phosphogypsum ($CaSO_4 \cdot 2\,H_2O$). Triple superphosphate (TSP) typically consists of 44-48% of P_2O_5 and no gypsum. A mixture of single superphosphate and triple superphosphate is called double superphosphate. More than 90% of a typical superphosphate fertilizer is water-soluble.

Multinutrient Fertilizers

These fertilizers are the most common. They consist of two or more nutrient components.

Binary (NP, NK, PK) Fertilizers

Major two-component fertilizers provide both nitrogen and phosphorus to the plants. These are called NP fertilizers. The main NP fertilizers are monoammonium phosphate (MAP) and diammonium phosphate (DAP). The active ingredient in MAP is $NH_4H_2PO_4$. The active ingredient in DAP is $(NH_4)_2HPO_4$. About 85% of MAP and DAP fertilizers are soluble in water.

NPK Fertilizers

NPK fertilizers are three-component fertilizers providing nitrogen, phosphorus, and potassium.

NPK rating is a rating system describing the amount of nitrogen, phosphorus, and potassium in a fertilizer. NPK ratings consist of three numbers separated by dashes (e.g., 10-10-10 or 16-4-8) describing the chemical content of fertilizers. The first number represents the percentage of nitrogen in the product; the second number, P_2O_5; the third, K_2O. Fertilizers do not actually contain P_2O_5 or K_2O, but the system is a conventional shorthand for the amount of the phosphorus (P) or potassium (K) in a fertilizer. A 50-pound (23 kg) bag of fertilizer labeled 16-4-8 contains 8 lb (3.6 kg) of nitrogen (16% of the 50 pounds), an amount of phosphorus equivalent to that in 2 pounds of P_2O_5 (4% of 50 pounds), and 4 pounds of K_2O (8% of 50 pounds). Most fertilizers are labeled according to this N-P-K convention, although Australian convention, following an N-P-K-S system, adds a fourth number for sulfur.

Micronutrients

The main micronutrients are molybdenum, zinc, and copper. These elements are provided as water-soluble salts. Iron presents special problems because it converts to insoluble (bio-unavailable) compounds at moderate soil pH and phosphate concentrations. For this reason, iron is often administered as a chelate complex, e.g., the EDTA derivative. The micronutrient needs depend on the plant. For example, sugar beets appear to require boron, and legumes require cobalt.

Production

Nitrogen Fertilizers

Top users of nitrogen-based fertilizer		
Country	**Total N use (Mt pa)**	**Amt. used for feed/pasture (Mt pa)**
China	18.7	3.0
India	11.9	N/A
U.S.	9.1	4.7
France	2.5	1.3

Germany	2.0	1.2
Brazil	1.7	0.7
Canada	1.6	0.9
Turkey	1.5	0.3
UK	1.3	0.9
Mexico	1.3	0.3
Spain	1.2	0.5
Argentina	0.4	0.1

Nitrogen fertilizers are made from ammonia (NH_3), which is sometimes injected into the ground directly. The ammonia is produced by the Haber-Bosch process. In this energy-intensive process, natural gas (CH_4) supplies the hydrogen, and the nitrogen (N_2) is derived from the air. This ammonia is used as a feedstock for all other nitrogen fertilizers, such as anhydrous ammonium nitrate (NH_4NO_3) and urea ($CO(NH_2)_2$).

Deposits of sodium nitrate ($NaNO_3$) (Chilean saltpeter) are also found in the Atacama desert in Chile and was one of the original (1830) nitrogen-rich fertilizers used. It is still mined for fertilizer.

There has been technical work investigating on-site (on-farm) synthesis of nitrate fertilizer using solar photovoltaic power, which would enable farmers more control in soil fertility, while using far less surface area than conventional organic farming for nitrogen fertilizer.

Phosphate Fertilizers

All phosphate fertilizers are obtained by extraction from minerals containing the anion PO_4^{3-}. In rare cases, fields are treated with the crushed mineral, but most often more soluble salts are produced by chemical treatment of phosphate minerals. The most popular phosphate-containing minerals are referred to collectively as phosphate rock. The main minerals are fluorapatite $Ca_5(PO_4)_3F$ (CFA) and hydroxyapatite $Ca_5(PO_4)_3OH$. These minerals are converted to water-soluble phosphate salts by treatment with sulfuric or phosphoric acids. The large production of sulfuric acid as an industrial chemical is primarily due to its use as cheap acid in processing phosphate rock into phosphate fertilizer. The global primary uses for both sulfur and phosphorus compounds relate to this basic process.

In the nitrophosphate process or Odda process (invented in 1927), phosphate rock with up to a 20% phosphorus (P) content is dissolved with nitric acid (HNO_3) to produce a mixture of phosphoric acid (H_3PO_4) and calcium nitrate ($Ca(NO_3)_2$). This mixture can be combined with a potassium fertilizer to produce a *compound fertilizer* with the three macronutrients N, P and K in easily dissolved form.

Potassium Fertilizers

Potash is a mixture of potassium minerals used to make potassium (chemical symbol: K) fertilizers. Potash is soluble in water, so the main effort in producing this nutrient from the ore involves some purification steps; e.g., to remove sodium chloride (NaCl) (common salt). Sometimes potash is referred to as K_2O, as a matter of convenience to those describing the potassium content. In fact potash fertilizers are usually potassium chloride, potassium sulfate, potassium carbonate, or potassium nitrate.

Compound Fertilizers

Compound fertilizers, which contain N, P, and K, can often be produced by mixing straight fertilizers. In some cases, chemical reactions occur between the two or more components. For example, monoammonium and diammonium phosphates, which provide plants with both N and P, are produced by neutralizing phosphoric acid (from phosphate rock) and ammonia :

$$NH_3 + H_3PO_4 \rightarrow (NH_4)H_2PO_4$$

$$2\ NH_3 + H_3PO_4 \rightarrow (NH_4)_2HPO_4$$

Organic Fertilizers

Compost bin for small-scale production of organic fertilizer

A large commercial compost operation

The main "organic fertilizers" are peat, animal wastes, plant wastes from agriculture, and treated sewage sludge (biosolids). In terms of volume, peat is the most widely used organic fertilizer. This immature form of coal confers no nutritional value to the plants, but improves the soil by aeration and absorbing water. Animal sources include the products of the slaughter of animals. Bloodmeal, bone meal, hides, hoofs, and horns are typical components. Organic fertilizer usually contain fewer nutrients, but offer other advantages as well as being appealing to those who are trying to practice "environmentally friendly" farming.

Other Elements: Calcium, Magnesium, and Sulfur

Calcium is supplied as superphosphate or calcium ammonium nitrate solutions.

Application

Fertilizers are commonly used for growing all crops, with application rates depending on the soil fertility, usually as measured by a soil test and according to the particular crop. Legumes, for example, fix nitrogen from the atmosphere and generally do not require nitrogen fertilizer.

Liquid vs Solid

Fertilizers are applied to crops both as solids and as liquid. About 90% of fertilizers are applied as solids. Solid fertilizer is typically granulated or powdered. Often solids are available as prills, a solid globule. Liquid fertilizers comprise anhydrous ammonia, aqueous solutions of ammonia, aqueous solutions of ammonium nitrate or urea. These concentrated products may be diluted with water to form a concentrated liquid fertilizer (e.g., UAN). Advantages of liquid fertilizer are its more rapid effect and easier coverage. The addition of fertilizer to irrigation water is called "fertigation".

Slow- and Controlled-release Fertilizers

Slow- and controlled-release involve only 0.15% (562,000 tons) of the fertilizer market (1995). Their utility stems from the fact that fertilizers are subject to antagonistic processes. In addition to their providing the nutrition to plants, excess fertilizers can be poisonous to the same plant. Competitive with the uptake by plants is the degradation or loss of the fertilizer. Microbes degrade many fertilizers, e.g., by immobilization or oxidation. Furthermore, fertilizers are lost by evaporation or leaching. Most slow-release fertilizers are derivatives of urea, a straight fertilizer providing nitrogen. Isobutylidenediurea ("IBDU") and urea-formaldehyde slowly convert in the soil to free urea, which is rapidly uptaken by plants. IBDU is a single compound with the formula $(CH_3)_2CHCH(NHC(O)NH_2)_2$ whereas the urea-formaldehydes consist of mixtures of the approximate formula $(HOCH_2NHC(O)NH)_nCH_2$.

Besides being more efficient in the utilization of the applied nutrients, slow-release technologies also reduce the impact on the environment and the contamination of the subsurface water. Slow-release fertilizers (various forms including fertilizer spikes, tabs, etc.) which reduce the problem of "burning" the plants due to excess nitrogen. Polymer coating of fertilizer ingredients gives tablets and spikes a 'true time-release' or 'staged nutrient release' (SNR) of fertilizer nutrients.

Controlled release fertilizers are traditional fertilizers encapsulated in a shell that degrades at a specified rate. Sulfur is a typical encapsulation material. Other coated products use thermoplastics (and sometimes ethylene-vinyl acetate and surfactants, etc.) to produce diffusion-controlled release of urea or other fertilizers. "Reactive Layer Coating" can produce thinner, hence cheaper, membrane coatings by applying reactive monomers simultaneously to the soluble particles. "Multicote" is a process applying layers of low-cost fatty acid salts with a paraffin topcoat.

Foliar Application

Foliar fertilizers are applied directly to leaves. The method is almost invariably used to apply water-soluble straight nitrogen fertilizers and used especially for high value crops such as fruits.

Fertilizer burn

Chemicals that Affect Nitrogen Uptake

Various chemicals are used to enhance the efficiency of nitrogen-based fertilizers. In this way farmers can limit the polluting effects of nitrogen run-off. Nitrification inhibitors (also known as nitrogen stabilizers) suppress the conversion of ammonia into nitrate, an anion that is more prone to leaching. 1-Carbamoyl-3-methylpyrazole (CMP), dicyandiamide, and nitrapyrin (2-chloro-6-trichloromethylpyridine) are popular. Urease inhibitors are used to slow the hydrolytic conversion of urea into ammonia, which is prone to evaporation as well as nitrification. The conversion of urea to ammonia catalyzed by enzymes called ureases. A popular inhibitor of ureases is N-(n-butyl)thiophosphoric triamide (NBPT).

Overfertilization

Careful fertilization technologies are important because excess nutrients can be as detrimental. Fertilizer burn can occur when too much fertilizer is applied, resulting in drying out of the leaves and damage or even death of the plant. Fertilizers vary in their tendency to burn roughly in accordance with their salt index.

Statistics

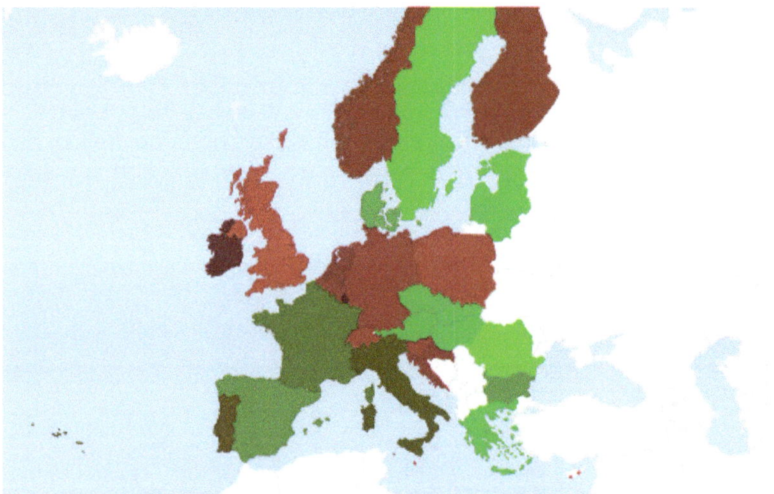

The map displays the statistics of fertilizer consumption in western and central European counties from data published by The World Bank for 2012.

Conservative estimates report 30 to 50% of crop yields are attributed to natural or synthetic commercial fertilizer. Global market value is likely to rise to more than US$185 billion until 2019. The European fertilizer market will grow to earn revenues of approx. €15.3 billion in 2018.

Data on the fertilizer consumption per hectare arable land in 2012 are published by The World Bank. For the diagram below values of the European Union (EU) countries have been extracted and are presented as kilograms per hectare (pounds per acre). The total consumption of fertilizer in the EU is 15.9 million tons for 105 million hectare arable land area (or 107 million hectare arable land according to another estimate). This figure equates to 151 kg of fertilizers consumed per ha arable land on average for the EU countries. Interestingly, mainly in those countries where fertilizers are consumed a lot also plant growth product are sold more than in others.

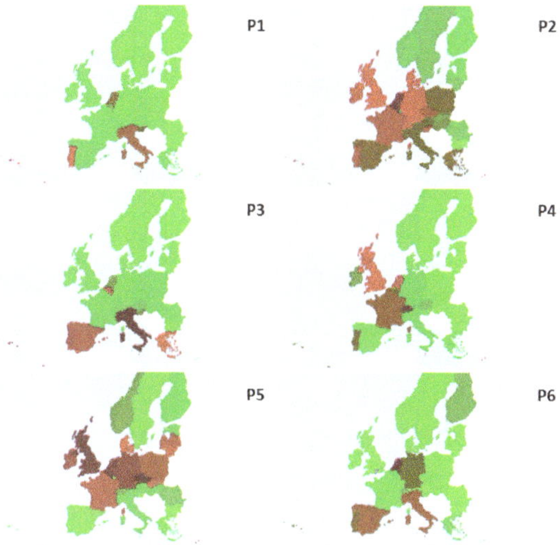

Pesticide categories, EUROSTAT. P5= Plant growth regulators. The red/green scale represents high/low pesticide sales per arable land.

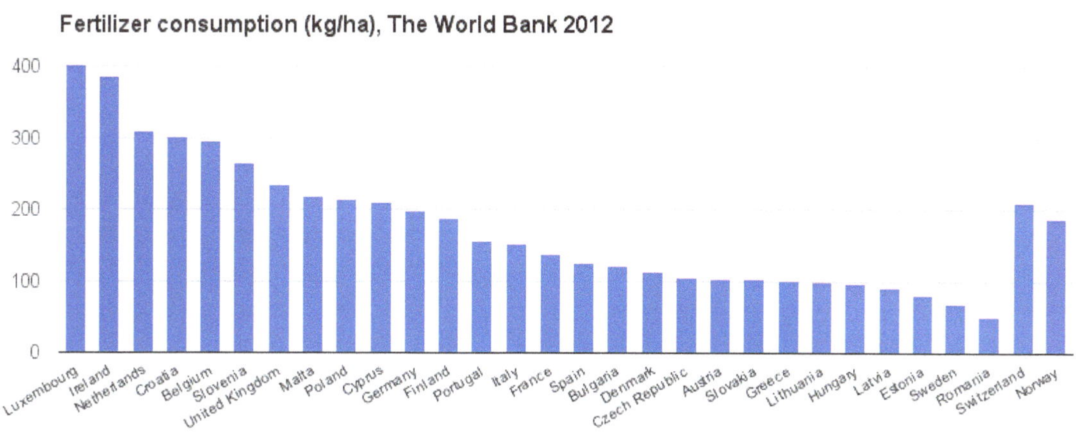

Environmental Effects

Runoff of soil and fertilizer during a rain storm

An algal bloom caused by eutrophication

Water

Agricultural run-off is a major contributor to the eutrophication of fresh water bodies. For example, in the US, about half of all the lakes are eutrophic. The main contributor to eutrophication is phosphate, which is normally a limiting nutrient; high concentrations promote the growth of cyanobacteria and algae, the demise of which consumes oxygen. Cyanobacteria blooms ('algal blooms') can also produce harmful toxins that can accumulate in the food chain, and can be harmful to humans.

The nitrogen-rich compounds found in fertilizer runoff are the primary cause of serious oxygen depletion in many parts of oceans, especially in coastal zones, lakes and rivers. The resulting lack of dissolved oxygen greatly reduces the ability of these areas to sustain oceanic fauna. The number of oceanic dead zones near inhabited coastlines are increasing. As of 2006, the application of nitrogen fertilizer is being increasingly controlled in northwestern Europe and the United States. If eutrophication *can* be reversed, it may take decades before the accumulated nitrates in groundwater can be broken down by natural processes.

Nitrate Pollution

Only a fraction of the nitrogen-based fertilizers is converted to produce and other plant matter. The remainder accumulates in the soil or lost as run-off. High application rates of nitrogen-containing fertilizers combined with the high water solubility of nitrate leads to increased runoff into surface water as well as leaching into groundwater, thereby causing groundwater pollution. The excessive use of nitrogen-containing fertilizers (be they synthetic or natural) is particularly damaging, as much of the nitrogen that is not taken up by plants is transformed into nitrate which is easily leached.

Nitrate levels above 10 mg/L (10 ppm) in groundwater can cause 'blue baby syndrome' (acquired methemoglobinemia). The nutrients, especially nitrates, in fertilizers can cause problems for natural habitats and for human health if they are washed off soil into watercourses or leached through soil into groundwater.

Soil

Acidification

Nitrogen-containing fertilizers can cause soil acidification when added. This may lead to decreases in nutrient availability which may be offset by liming.

Accumulation of Toxic Elements

Cadmium

The concentration of cadmium in phosphorus-containing fertilizers varies considerably and can be problematic. For example, mono-ammonium phosphate fertilizer may have a cadmium content of as low as 0.14 mg/kg or as high as 50.9 mg/kg. This is because the phosphate rock used in their manufacture can contain as much as 188 mg/kg cadmium (examples are deposits on Nauru and the Christmas islands). Continuous use of high-cadmium fertilizer can contaminate soil and plants. Limits to the cadmium content of phosphate fertilizers has been considered by the European Commission. Producers of phosphorus-containing fertilizers now select phosphate rock based on the cadmium content.

Fluoride

Phosphate rocks contain high levels of fluoride. Consequently, the widespread use of phosphate fertilizers has increased soil fluoride concentrations. It has been found that food contamination from fertilizer is of little concern as plants accumulate little fluoride from the soil; of greater concern is the possibility of fluoride toxicity to livestock that ingest contaminated soils. Also of possible concern are the effects of fluoride on soil microorganisms.

Radioactive Elements

The radioactive content of the fertilizers varies considerably and depends both on their concentrations in the parent mineral and on the fertilizer production process. Uranium-238 concentrations range can range from 7 to 100 pCi/g in phosphate rock and from 1 to 67 pCi/g in phosphate fertilizers. Where high annual rates of phosphorus fertilizer are used, this can result in uranium-238 concentrations in soils and drainage waters that are several times greater than are normally present. However, the impact of these increases on the risk to human health from radinuclide contamination of foods is very small (less than 0.05 mSv/y).

Other Metals

Steel industry wastes, recycled into fertilizers for their high levels of zinc (essential to plant growth), wastes can include the following toxic metals: lead arsenic, cadmium, chromium, and nickel. The most common toxic elements in this type of fertilizer are mercury, lead, and arsenic. These potentially harmful impurities can be removed; however, this significantly increases cost. Highly pure fertilizers are widely available and perhaps best known as the highly water-soluble fertilizers containing blue dyes used around households, such as Miracle-Gro. These highly water-soluble fertilizers are used in the plant nursery business and are available in larger packages at significantly less cost than retail quantities. There are also some inexpensive retail granular garden fertilizers made with high purity ingredients.

Trace Mineral Depletion

Attention has been addressed to the decreasing concentrations of elements such as iron, zinc, copper and magnesium in many foods over the last 50–60 years. Intensive farming practices, including the use of synthetic fertilizers are frequently suggested as reasons for these declines and organic farming is often suggested as a solution. Although improved crop yields resulting from NPK fertilizers are known to dilute the concentrations of other nutrients in plants, much of the measured decline can be attributed to the use of progressively higher-yielding crop varieties which produce foods with lower mineral concentrations than their less productive ancestors. It is, therefore, unlikely that organic farming or reduced use of fertilizers will solve the problem; foods with high nutrient density are posited to be achieved using older, lower-yielding varieties or the development of new high-yield, nutrient-dense varieties.

Fertilizers are, in fact, more likely to solve trace mineral deficiency problems than cause them: In Western Australia deficiencies of zinc, copper, manganese, iron and molybdenum were identified as limiting the growth of broad-acre crops and pastures in the 1940s and 1950s. Soils in Western Australia are very old, highly weathered and deficient in many of the major nutrients and trace elements. Since this time these trace elements are routinely added to fertilizers used in agriculture in this state. Many other soils around the world are deficient in zinc, leading to deficiency in both plants and humans, and zinc fertilizers are widely used to solve this problem.

Changes in Soil Biology

High levels of fertilizer may cause the breakdown of the symbiotic relationships between plant roots and mycorrhizal fungi.

Energy Consumption and Sustainability

In the USA in 2004, 317 billion cubic feet of natural gas were consumed in the industrial production of ammonia, less than 1.5% of total U.S. annual consumption of natural gas. A 2002 report suggested that the production of ammonia consumes about 5% of global natural gas consumption, which is somewhat under 2% of world energy production.

Ammonia is produced from natural gas and air. The cost of natural gas makes up about 90% of the cost of producing ammonia. The increase in price of natural gases over the past decade, along with other factors such as increasing demand, have contributed to an increase in fertilizer price.

Contribution to Climate Change

The greenhouse gases carbon dioxide, methane and nitrous oxide are produced during the manufacture of nitrogen fertilizer. The effects can be combined into an equivalent amount of carbon dioxide. The amount varies according to the efficiency of the process. The figure for the United Kingdom is over 2 kilogrammes of carbon dioxide equivalent for each kilogramme of ammonium nitrate. Nitrogen fertilizer can be converted by soil bacteria to nitrous oxide, a greenhouse gas.

Atmosphere

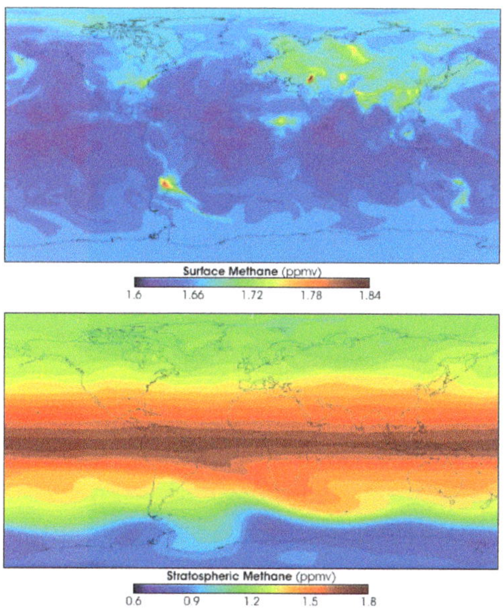

Global methane concentrations (surface and atmospheric) for 2005; note distinct plumes

Through the increasing use of nitrogen fertilizer, which was used at a rate of about 110 million tons (of N) per year in 2012, adding to the already existing amount of reactive nitrogen, nitrous oxide (N_2O) has become the third most important greenhouse gas after carbon dioxide and methane. It has a global warming potential 296 times larger than an equal mass of carbon dioxide and it also contributes to stratospheric ozone depletion. By changing processes and procedures, it is possible to mitigate some, but not all, of these effects on anthropogenic climate change.

Methane emissions from crop fields (notably rice paddy fields) are increased by the application of ammonium-based fertilizers. These emissions contribute to global climate change as methane is a potent greenhouse gas.

Regulation

In Europe problems with high nitrate concentrations in run-off are being addressed by the European Union's Nitrates Directive. Within Britain, farmers are encouraged to manage their land more sustainably in 'catchment-sensitive farming'. In the US, high concentrations of nitrate and phosphorus in runoff and drainage water are classified as non-point source pollutants due to their diffuse origin; this pollution is regulated at state level. Oregon and Washington, both in the United States, have fertilizer registration programs with on-line databases listing chemical analyses of fertilizers.

History

Founded in 1812, Mirat, producer of manures and fertilizers, is claimed to be the oldest industrial business in Salamanca (Spain).

Management of soil fertility has been the preoccupation of farmers for thousands of years. Egyptians, Romans, Babylonians, and early Germans all are recorded as using minerals and or manure to enhance the productivity of their farms. The modern science of plant nutrition started in the 19th century and the work of German chemist Justus von Liebig, among others. John Bennet Lawes, an English entrepreneur, began to experiment on the effects of various manures on plants growing in pots in 1837, and a year or two later the experiments were extended to crops in the field. One immediate consequence was that in 1842 he patented a manure formed by treating phosphates with sulphuric acid, and thus was the first to create the artificial manure industry. In the succeeding year he enlisted the services of Joseph Henry Gilbert, with whom he carried on for more than half a century on experiments in raising crops at the Institute of Arable Crops Research.

The Birkeland–Eyde process was one of the competing industrial processes in the beginning of nitrogen based fertilizer production. This process was used to fix atmospheric nitrogen (N_2) into nitric acid (HNO_3), one of several chemical processes generally referred to as nitrogen fixation. The resultant nitric acid was then used as a source of nitrate (NO_3^-). A factory based on the process was built in Rjukan and Notodden in Norway, combined with the building of large hydroelectric power facilities.

The 1910s and 1920s witness the rise of the Haber process and the Ostwald process. The Haber process produces ammonia (NH_3) from methane (CH_4) gas and molecular nitrogen (N_2). The ammonia from the Haber process is then converted into nitric acid (HNO_3) in the Ostwald process. The development of synthetic fertilizer has significantly supported global population growth — it has been estimated that almost half the people on the Earth are currently fed as a result of synthetic nitrogen fertilizer use.

The use of commercial fertilizers has increased steadily in the last 50 years, rising almost 20-fold to the current rate of 100 million tonnes of nitrogen per year. Without commercial fertilizers it is estimated that about one-third of the food produced now could not be produced. The use of phosphate fertilizers has also increased from 9 million tonnes per year in 1960 to 40 million tonnes per year in 2000. A maize crop yielding 6–9 tonnes of grain per hectare (2.5 acres) requires 31–50 kilograms (68–110 lb) of phosphate fertilizer to be applied; soybean crops require about half, as 20–25 kg per hectare. Yara International is the world's largest producer of nitrogen-based fertilizers.

Controlled-nitrogen-release technologies based on polymers derived from combining urea and formaldehyde were first produced in 1936 and commercialized in 1955. The early product had 60 percent of the total nitrogen cold-water-insoluble, and the unreacted (quick-release) less than 15%. Methylene ureas were commercialized in the 1960s and 1970s, having 25% and 60% of the nitrogen as cold-water-insoluble, and unreacted urea nitrogen in the range of 15% to 30%.

In the 1960s, the Tennessee Valley Authority National Fertilizer Development Center began developing sulfur-coated urea; sulfur was used as the principal coating material because of its low cost and its value as a secondary nutrient. Usually there is another wax or polymer which seals the sulfur; the slow-release properties depend on the degradation of the secondary sealant by soil microbes as well as mechanical imperfections (cracks, etc.) in the sulfur. They typically provide 6 to 16 weeks of delayed release in turf applications. When a hard polymer is used as the secondary coating, the properties are a cross between diffusion-controlled particles and traditional sulfur-coated.

Labeling of Fertilizer

The labeling of fertilizers varies by country in terms of analysis methodology, nutrient labeling, and minimum nutrient requirements. The most common labeling convention shows the amounts of nitrogen, phosphorus, and potassium in the fertilizer.

Labeling of Macronutrient Fertilizers

Macronutrient fertilizers are generally labeled with an *NPK* analysis, based on the relative content of the chemical elements nitrogen (N), phosphorus (P), and potassium (K) that are commonly used in fertilizers. However, numbers used in this labeling scheme do not directly represent the source composition or absolute nutrient content of the fertilizer. The N value is the percentage of elemental nitrogen by weight in the fertilizer. The value for P is the fraction by weight of P_2O_5 in a fertilizer with the same amount of phosphorus that gets all of its phosphorus from P_2O_5. The value for K is analogous, based on a fertilizer with K_2O.

For example, the fertilizer *sylvite* is a naturally occurring mineral consisting mostly of potassium chloride (*KCl*). As such, it contains one potassium atom for every chlorine atom, and is 52% potassium and 48% chlorine *by weight* (because of the difference in atomic weights of the elements). K_2O is similarly 83% potassium. Therefore, a fertilizer that gets all its potassium from KCI would have to be 63% K_2O (.52/.83 is .63). Pure KCl fertilizer would thus be labeled 0-0-63; because sylvite is less than pure (it contains other compounds that contain no potassium), it is labeled 0-0-60.

Converting Nutrient Analysis to Composition

The factors for converting from P_2O_5 and K_2O values on a fertilizer label to the concentrations (by weight) of P and K elements are as follows:

- P_2O_5 consists of 56.4% oxygen and 43.6% elemental phosphorus. The percentage (mass fraction) of elemental phosphorus is 43.6% so elemental P = 0.436 x P_2O_5
- K_2O consists of 17% oxygen and 83% elemental potassium. The percentage (mass fraction) of elemental potassium is 83% so elemental K = 0.83 x K_2O
- Nitrogen values represent actual nitrogen content so these numbers do not need to be converted.

Using these conversion factors, an 18–51–20 fertilizer contains by weight:

- 18% elemental (N)
- 22% elemental (P), and
- 17% elemental (K)

Other Labeling Conventions

In the U.K., fertilizer labeling regulations allow for reporting the elemental mass fractions of phosphorus and potassium. The regulations stipulate that this should be done in parentheses after the standard N-P-K values. In Australia, macronutrient fertilizers are labeled with an "N-P-K-S" system, which uses elemental mass fractions rather than standard N-P-K values and includes the amount of sulfur (S) contained in the fertilizer.

NPK Values for Commercial Fertilizers

NPK Values for Various Synthetic Fertilizers

- 15-00-00 Calcium nitrate
- 21-00-00 Ammonium sulfate
- 30-00-00 to 40-00-00 Sulfur-coated urea (slow release)
- 31-00-00 Isobutylidene diurea (~90% slow release)
- 33-00-00 to 34-00-00 Ammonium nitrate
- 35-00-00 Ureaform (~85% slow release, sparingly soluble ureaformaldehyde)
- 40-00-00 Methylene ureas (~70% slow release)
- 46-00-00 Urea (U-46)
- 82-00-00 Anhydrous ammonia
- 10-34-00 to 11-37-00 Ammonium polyphosphate
- 11-48-00 to 11-55-00 Monoammonium phosphate
- 18-46-00 to 21-54-00 Diammonium phosphate
- 13-00-44 Potassium nitrate
- 00-17-00 to 00-22-00 Superphosphate (Monocalcium phosphate monohydrate with gypsum)
- 00-44-00 to 00-52-00 Triple superphosphate (Monocalcium phosphate monohydrate)

NPK Values for Mined Fertilizer Minerals

- 11-08-02 to 16-12-03 bird guano
- 00-3-00 to 00-8-00 Raw Phosphate Rock (would be 00-34-00 if it were soluble)
- 00-00-22 Potassium magnesium sulfate (K-mag)
- 00-00-60 Potassium chloride

NPK Values for Biosolids Fertilizers and Others

- 09-00-00 dairy manure

- 01-00-01 horse manure
- 03-02-02 poultry manure
- 04-12-00 Bone meal
- 05-05-06 Fish blood and bone
- 06-02-00 Milorganite

Soil Chemistry

Soil chemistry is the study of the chemical characteristics of soil. Soil chemistry is affected by mineral composition, organic matter and environmental factors.

Soil chemistry is the study of the chemical characteristics of soil. Soil chemistry is affected by mineral composition, organic matter and environmental factors. Soil - (i) The unconsolidated mineral or organic material on the immediate surface of the Earth that serves as a natural medium for the growth of land plants. (ii) The unconsolidated mineral or organic matter on the surface of the Earth that has been subjected to and shows effects of genetic and environmental factors of: climate (including water and temperature effects), and macro- and micro organisms, conditioned by relief, acting on parent material over a period of time. A product-soil differs from the material from which it is derived in many physical, chemical,biological, and morphological properties and characteristics. This definition is from Soil Taxonomy, second edition.

Key facts you should know about Soil:

Soil makes up the outermost layer of our planet and is formed from rocks and decaying plants and Animals. Soil has varying amounts of organic matter (living and dead organisms), minerals, and nutrients. An average soil sample is 45 percent minerals, 25 percent water, 25 percent air, and five percent organic matter. Different-sized mineral particles, such as sand, silt, and clay, give soil its texture. Topsoil is the most productive soil layer. Ten tonnes of topsoil spread evenly over a hectare is only as thick as a one Euro coin. Natural processes can take more than 500 years to form 2 centimeters of topsoil. In some cases, 5 tonnes of animal life can live in one hectare of soil. Fungi and bacteria help break down organic matter in the soil. Earthworms digest organic matter, recycle nutrients, and make the surface soil richer. Roots loosen the soil, allowing oxygen to penetrate. This benefits animals living in the soil. They also benefit roots which require oxygen themselves. They further hold soil together and help prevent erosion. A fully functioning soil reduces the risk of floods and protects underground water supplies by neutralizing or filtering out potential pollutants and storing as much as 3750 tonnes of water per hectare. Scientists have found 10,000 types of soil in Europe and about 70,000 types of soil in the United States. Soil stores 10% of the world's carbon dioxide emissions.

Environmental Soil Chemistry

A knowledge of environmental soil chemistry is paramount to predicting the fate of contaminants, as well as the processes by which they are initially released into the soil. Once a chemical is exposed to the soil environment myriad chemical reactions can occur that may increase or decrease contaminant toxicity. These reactions include adsorption/desorption, precipitation, polymerization, dissolution, complexation and oxidation/reduction. These reactions are often disregarded by scientists and engineers involved with environmental remediation. Understanding these processes enable us to better predict the fate and toxicity of contaminants and provide the knowledge to develop scientifically correct, and cost-effective remediation strategies.

Concepts

- Anion and cation exchange capacity
- Soil pH
- Mineral formation and transformation processes
- Clay mineralogy
- Sorption and precipitation reactions in soil
- Oxidation-reduction reactions
- Chemistry of problem soils

Clay Minerals

A soil particle may be a mineral or a rock fragment. A mineral is a chemical compound formed in nature during a geological process, whereas a rock fragment has a combination of one or more minerals. Based on the nature of atoms, minerals are classified as silicates, aluminates, oxides, carbonates and phosphates.

Out of these, silicate minerals are the most important as they influence the properties of clay soils. Different arrangements of atoms in the silicate minerals give rise to different silicate structures.

Basic Structural Units

Soil minerals are formed from two basic structural units: tetrahedral and octahedral. Considering the valencies of the atoms forming the units, it is clear that the units are not electrically neutral and as such do not exist as single units.

The basic units combine to form sheets in which the oxygen or hydroxyl ions are shared among adjacent units. Three types of sheets are thus formed, namely *silica sheet, gibbsite sheet* and *brucite sheet.*

Isomorphous substitution is the replacement of the central atom of the tetrahedral or octahedral unit by another atom during the formation of the sheets.

The sheets then combine to form various two-layer or three-layer sheet minerals. As the basic units of clay minerals are sheet-like structures, the particle formed from stacking of the basic units is also plate-like. As a result, the surface area per unit mass becomes very large.

Structure of Clay Minerals

A tetrahedral unit consists of a central silicon atom that is surrounded by four oxygen atoms located at the corners of a tetrahedron. A combination of tetrahedrons forms a *silica sheet.*

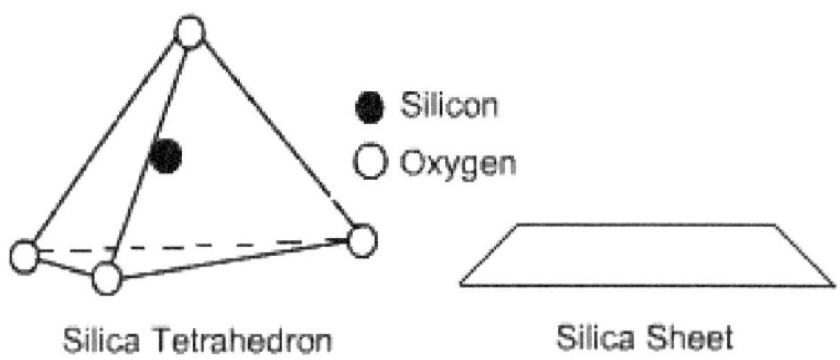

An octahedral unit consists of a central ion, either aluminium or magnesium, that is surrounded by six hydroxyl ions located at the corners of an octahedron. A combination of aluminium-hydroxyl octahedrons forms a *gibbsite sheet*, whereas a combination of magnesium-hydroxyl octahedrons forms a *brucite sheet.*

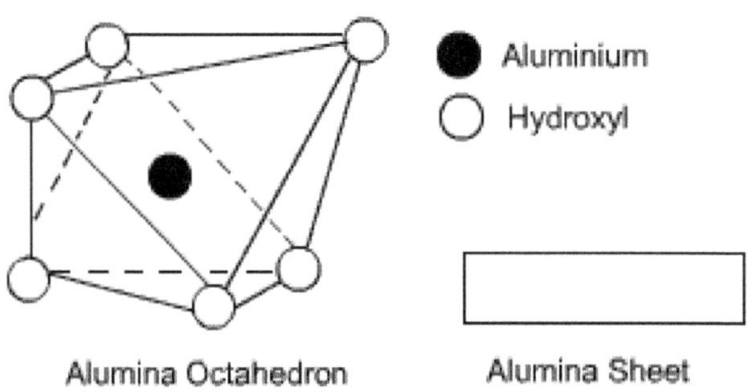

Two-layer Sheet Minerals

Kaolinite and halloysite clay minerals are the most common.

Kaolinite Mineral

The basic kaolinite unit is a two-layer unit that is formed by stacking a gibbsite sheet on a silica sheet. These basic units are then stacked one on top of the other to form a lattice of the mineral. The units are held together by hydrogen bonds. The strong bonding does not permit water to enter the lattice. Thus, kaolinite minerals are stable and do not expand under saturation.

Kaolinite is the most abundant constituent of residual clay deposits.

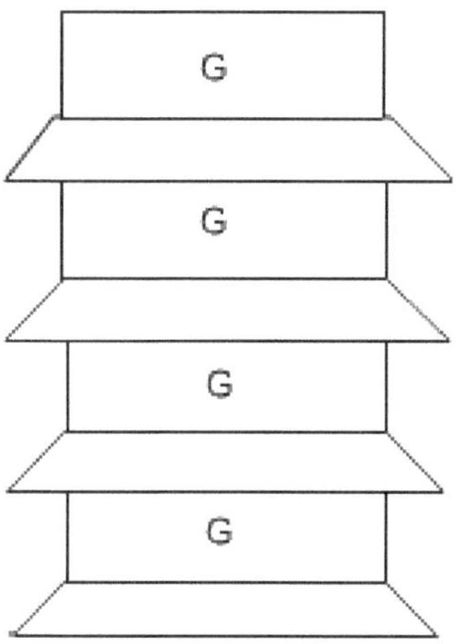

Fig: Kaolinite Mineral

Halloysite Mineral

The basic unit is also a two-layer sheet similar to that of kaolinite except for the presence of water between the sheets.

Three-layer Sheet Minerals

Montmorillonite and illite clay minerals are the most common. A basic three-layer sheet unit is formed by keeping one silica sheet each on the top and at the bottom of a gibbsite sheet. These units are stacked to form a lattice as shown.

Montmorillonite Mineral

The bonding between the three-layer units is by van der Waals forces. This bonding is very weak and water can enter easily. Thus, this mineral can imbibe a large quantity of water causing swelling. During dry weather, there will be shrinkage.

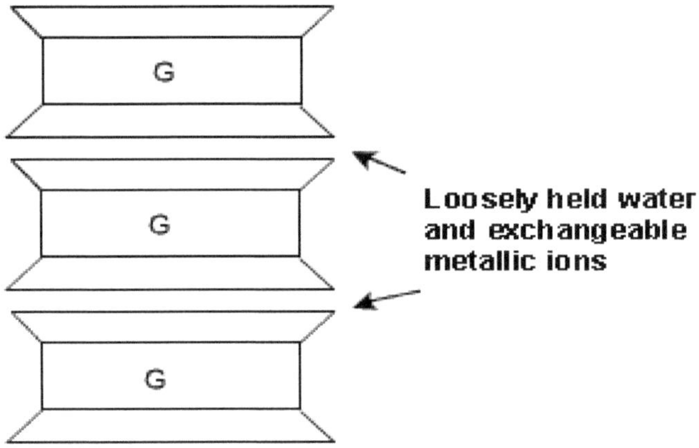

Fig:Montmorillonite Mineral

Illite Mineral

Illite consists of the basic montmorillonite units but are bonded by secondary valence forces and potassium ions, as shown. There is about 20% replacement of aluminium with silicon in the gibbsite sheet due to *isomorphous substitution*. This mineral is very stable and does not swell or shrink.

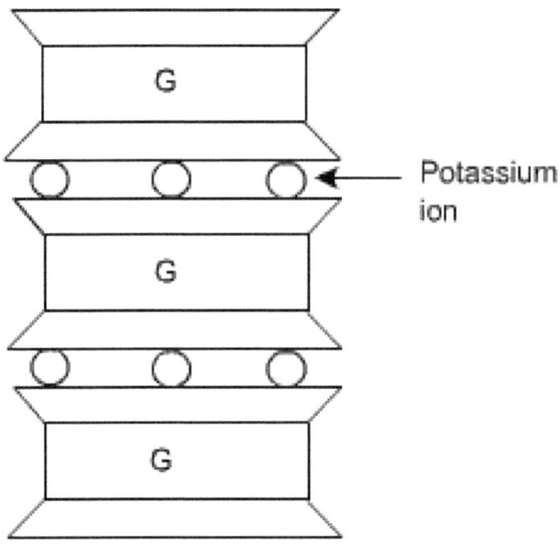

Fig: Illite Mineral

Fine Soil Fabric

Natural soils are rarely the same from one point in the ground to another. The content and nature of grains varies, but more importantly, so does the arrangement of these. The arrangement and organisation of particles and other features within a soil mass is termed its fabric.

CLAY particles are flaky. Their thickness is very small relative to their length & breadth, in some cases as thin as 1/100th of the length. They therefore have high specific surface values. These surfaces carry negative electrical charge, which attracts positive ions present in the pore water. Thus a lot of water may be held as adsorbed water within a clay mass.

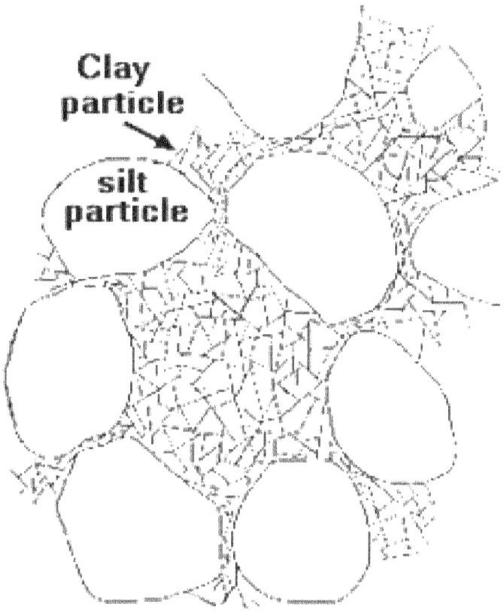

References

- Stewart, W.M.; Dibb, D.W.; Johnston, A.E.; Smyth, T.J. (2005). "The Contribution of Commercial Fertilizer Nutrients to Food Production". Agronomy Journal. 97: 1–6. doi:10.2134/agronj2005.0001
- Santamarina, J. Carlos, Katherine A. Klein, & Moheb A. Fam. Soils and Waves: Particulate Materials Behavior, Characterization and Process Monitoring. Wiley, 2001; pp. 35-36 & 51-53. ISBN 978-0-471-49058-6
- Khater, A. E. M. (2008). "Uranium and heavy metals in phosphate fertilizers" (PDF). www.radioecology.info. Retrieved 17 July 2014
- Soil Science Glossary Terms Committee (2008). Glossary of Soil Science Terms 2008. Madison, WI: Soil Science Society of America. ISBN 978-0-89118-851-3
- "Label Requirements of specialty and other bagged fertilizers". Michigan Department of Agriculture and Rural Development. Retrieved 14 March 2013
- "Phosphorus Solubility in Response to Acidification of Dairy Manure Amended Soils". Soil Science Society of America Journal. 72: 238. doi:10.2136/sssaj2007.0071N
- "National Code of Practice for Fertilizer Description & Labelling" (PDF). Australian Government Department of Agriculture, Fisheries and Forestry. Retrieved 14 March 2013
- Chesworth, Ward (2008). Encyclopedia of soil science. Dordrecht, Netherlands: Springer. p. 694. ISBN 1402039948. Retrieved 2 July 2016
- Davis, D.R.; Epp, M.D.; Riordan, H.D. (2004). "Changes in USDA Food Composition Data for 43 Garden Crops, 1950 to 1999". Journal of the American College of Nutrition. 23: 669–682. doi:10.1080/07315724.2004.10719409

- F. Stuart Chapin III; Pamela A. Matson; Harold A. Moon (2002). Principles of Terrestrial Ecosystem Ecology. Springer. ISBN 0387954392
- Livestock's Long Shadow: Environmental Issues and Options, Table 3.3. Retrieved 29 June 2009. United Nations Food and Agriculture Organization
- Roy, R. N.; Misra, R. V.; Montanez, A. (2002). "Decreasing reliance on mineral nitrogen-yet more food" (PDF). AMBIO: A Journal of the Human Environment. 31 (2): 177–183. doi:10.1579/0044-7447-31.2.177. Retrieved 3 July 2014
- Kötke, William H. (1993). The Final Empire: The Collapse of Civilization and the Seed of the Future. Arrow Point Press. ISBN 0963378457
- "A Farmer's Guide To Agriculture and Water Quality Issues: 3. Environmental Requirements & Incentive Programs For Nutrient Management". www.cals.ncsu.edu. Retrieved 3 July 2014
- J. Benton Jones, Jr. "Inorganic Chemical Fertilizers and Their Properties" in Plant Nutrition and Soil Fertility Manual, Second Edition. CRC Press, 2012. ISBN 978-1-4398-1609-7. eBook ISBN 978-1-4398-1610-3
- State-EPA Nutrient Innovations Task Group (2009). "An Urgent Call to Action – Report of the State-EPA Nutrient Innovations Task Group" (PDF). water.epa.gov. Retrieved 3 July 2014
- Moore, Geoff (2001). Soilguide - A handbook for understanding and managing agricultural soils (PDF). Perth, Western Australia: Agriculture Western Australia. pp. 161–207. ISBN 0 7307 0057 7
- C. J. Rosen; B. P. Horgan (9 January 2009). "Preventing Pollution Problems from Lawn and Garden Fertilizers". Extension.umn.edu. Retrieved 25 August 2010
- Carroll and Salt, Steven B. and Steven D. (2004). Ecology for Gardeners. Cambridge: Timber Press. ISBN 9780881926118
- "NOFA Interstate Council: The Natural Farmer. Ecologically Sound Nitrogen Management. Mark Schonbeck". Nofa.org. 25 February 2004. Retrieved 25 August 2010
- Appl, Max (2000). Ullmann's Encyclopedia of Industrial Chemistry, Volume 3. Weinheim, Germany: Wiley-VCH Verlag GmbH & Co. KGaA. pp. 139–225. doi:10.1002/14356007.002_011. ISBN 9783527306732
- Lugon-Moulin, N.; Ryan, L.; Donini, P.; Rossi, L. (2006). "Cadmium content of phosphate fertilizers used for tobacco production" (PDF). Agron. Sustain. Dev. 26: 151–155. doi:10.1051/agro:2006010. Retrieved 27 June 2014
- G. J. Leigh (2004). The world's greatest fix: a history of nitrogen and agriculture. Oxford University Press US. pp. 134–139. ISBN 0-19-516582-9
- Oosterhuis, F.H.; Brouwer, F.M.; Wijnants, H.J. (2000). "A possible EU wide charge on cadmium in phosphate fertilisers: Economic and environmental implications." (PDF). dare.ubvu.vu.nl. Retrieved 27 June 2014
- Trevor Illtyd Williams; Thomas Kingston Derry (1982). A short history of twentieth-century technology c. 1900-c. 1950. Oxford University Press. pp. 134–135. ISBN 0-19-858159-9

3

Land Retrogression and Degradation

The major components of pressure on land have been discussed. Total vertical stress on a ground surface is the load applied to soil, which gets carried by solid grains and water. Also explored are the topics like soil retrogression and degradation, land degradation, shear strength, etc. This chapter provides a plethora of interdisciplinary topics for better comprehension of soil mechanics.

Soil Retrogression and Degradation

Soil retrogression and degradation are two regressive evolution processes associated with the loss of equilibrium of a stable soil. Retrogression is primarily due to soil erosion and corresponds to a phenomenon where succession reverts the land to its natural physical state. Degradation is an evolution, different from natural evolution, related to the local climate and vegetation. It is due to the replacement of primary plant communities (known as climax vegetation) by the secondary communities. This replacement modifies the humus composition and amount, and affects the formation of the soil. It is directly related to human activity. Soil degradation may also be viewed as any change or ecological disturbance to the soil perceived to be deleterious or undesirable.

General

At the beginning of soil formation, the bare rock out crops is gradually colonized by pioneer species (lichens and mosses). They are succeeded by herbaceous vegetation, shrubs and finally forest. In parallel, the first humus-bearing horizon is formed (the A horizon), followed by some mineral horizons (B horizons). Each successive stage is characterized by a certain association of soil/vegetation and environment, which defines an ecosystem.

After a certain time of parallel evolution between the ground and the vegetation, a state of steady balance is reached. This stage of development is called climax by some ecologists and "natural potential" by others. Succession is the evolution towards climax. Regardless of its name, the equilibrium stage of primary succession is the highest natural form of development that the environmental factors are capable of producing.

The cycles of evolution of soils have very variable durations, between tens, hundreds, or thousands of years for quickly evolving soils (A horizon only) to more than a million years for slowly developing soils. The same soil may achieve several successive steady state conditions during its existence, as exhibited by the Pygmy forest sequence in Mendocino County, California. Soils naturally reach a state of high productivity, from which they naturally degrade as mineral nutrients are removed from the soil system. Thus older soils are more vulnerable to the effects of induced retrogression and degradation.

Intensive tillage result on soil degradation.

Willow hedge strengthened with fascines for the limitation of runoff, north of France.

Ecological Factors Influencing Soil Formation

There are two types of ecological factors influencing the evolution of a soil (through alteration and humification). These two factors are extremely significant to explain the evolution of soils of short development.

- A first type of factor is the average climate of an area and the vegetation which is associated (biome).
- A second type of factor is more local, and is related to the original rock and local drainage. This type of factor explains appearance of specialized associations (ex peat bogs).

Biorhexistasy Theory

The destruction of the vegetation implies the destruction of evoluted soils, or a regressive evolution. Cycles of succession-regression of soils follow one another within short intervals of time (human actions) or long intervals of time (climate variations).

The climate role in the deterioration of the rocks and the formation of soils lead to the formulation of the theory of the biorhexistasy.

- In wet climate, the conditions are favorable to the deterioration of the rocks (mostly chemically), the development of the vegetation and the formation of soils; this period favorable to life is called biostasy.
- In dry climate, the rocks exposed are mostly subjected to mechanical disintegration which produces coarse detrital materials: this is referred to as rhexistasy.

Perturbations of the Balance of a Soil

When the state of balance, characterized by the ecosystem climax is reached, it tends to be maintained stable in the course of time. The vegetation installed on the ground provides the humus and ensures the ascending circulation of the matters. It protects the ground from erosion by playing the role of barrier (for example, protection from water and wind). Plants can also reduce erosion by binding the particles of the ground to their roots.

A disturbance of climax will cause retrogression, but often, secondary succession will start to guide the evolution of the system after that disturbance. Secondary succession is much faster than primary because the soil is already formed, although deteriorated and needing restoration as well.

However, when a significant destruction of the vegetation takes place (of natural origin such as an avalanche or human origin), the disturbance undergone by the ecosystem is too important. In this latter case, erosion is responsible for the destruction of the upper horizons of the ground, and is at the origin of a phenomenon of reversion to pioneer conditions. The phenomenon is called retrogression and can be partial or total (in this case, nothing remains beside bare rock). For example, the clearing of an inclined ground, subjected to violent rains, can lead to the complete destruction of the soil. Man can deeply modify the evolution of the soils by direct and brutal action, such as clearing, abusive cuts, forest pasture, litters raking. The climax vegetation is gradually replaced and the soil modified (example: replacement of leafy tree forests by moors or pines plantations). Retrogression is often related to very old human practices.

Influence of Human Activity

Soil erosion is the main factor for soil degradation and is due to several mechanisms: water erosion, wind erosion, chemical degradation and physical degradation.

Erosion is strongly related to human activity. For example, roads which increase impermeable surfaces lead to streaming and ground loss. Agriculture also accelerates soil erosion (increase of field size, correlated to hedges and ditches removal). Meadows are in regression to the profit of plowed lands. Spring cultures (sunflower, corn, beet) surfaces are increasing and leave the ground naked in winter. Sloping grounds are gradually colonized by vine. Lastly, use of herbicides leaves the ground naked between each crop. New cultural practices, such as mechanization also increases the risks of erosion. Fertilization by mineral manures rather than organic manure gradually destructure the soil. Many scientistsobserved a gradual decrease of soil organic matter content in soils, as well as a decrease of soil biological activity (in particular, in relation to chemical uses). Lastly, deforestation, in particular, is responsible for degradation of forest soils.

Agriculture increases the risk of erosion through its disturbance of vegetation by way of:

- Overgrazing of animals
- Monoculture planting
- Row cropping
- Tilling or plowing
- Crop removal
- Land-use conversion

Consequences of Soil Regression and Degradation

- yields impact: Recent increases in the human population have placed a great strain on the world's soil systems. More than 6 billion people are now using about 38% of the land area of the Earth to raise crops and livestock. Many soils suffer from various types of degradation, that can ultimately reduce their ability to produce food resources. Slight degradation refers to land where yield potential has been reduced by 10%, moderate degradation refers to a yield decrease from 10-50%. Severely degraded soils have lost more than 50% of their potential. Most severely degraded soils are located in developing countries.

- natural disasters: natural disasters such as mud flows, floods are responsible for the death of many living beings each year.

- deterioration of the water quality: the increase in the turbidity of water and the contribution of nitrogen and of phosphorus can result in eutrophication. Soils particles in surface waters are also accompanied by agricultural inputs and by some pollutants of industrial, urban and road origin (such as heavy metals). The ecological impact of agricultural inputs (such as weed killer) is known but difficult to evaluate because of the multiplicity of the products and their broad spectrum of action.

- biological diversity: soil degradation may involve perturbation of microbial communities, disappearance of the climax vegetation and decrease in animal habitat, thus leading to a biodiversity loss and animal extinction.

- economic loss: the estimated costs for land degradation are US $40 billion per year. This figure does not take into account negative externalities such as the increased use of fertilizers, loss of biodiversity and loss of unique landscapes.

Soil Enhancement, Rebuilding, and Regeneration

Problems of soil erosion can be fought, and certain practices can lead to soil enhancement and rebuilding. Even though simple, methods for reducing erosion are often not chosen because these practices outweigh the short-term benefits. Rebuilding is especially possible through the improvement of soil structure, addition of organic matter and limitation of runoff. However, these techniques will never totally succeed to restore a soil (and the fauna and flora associated to it) that took more than 1000 years to build up. Soil regeneration is the reformation of degraded soil through biological, chemical, and or physical processes.

When productivity declined in the low-clay soils of northern Thailand, farmers initially responded by adding organic matter from termite mounds, but this was unsustainable in the long-term. Scientists experimented with adding bentonite, one of the smectite family of clays, to the soil. In field trials, conducted by scientists from the International Water Management Institute in cooperation with Khon Kaen University and local farmers, this had the effect of helping retain water and nutrients. Supplementing the farmer's usual practice with a single application of 200 kg bentonite per rai (6.26 rai = 1 hectare) resulted in an average yield increase of 73%. More work showed that applying bentonite to degraded sandy soils reduced the risk of crop failure during drought years.

In 2008, three years after the initial trials, IWMI scientists conducted a survey among 250 farmers in northeast Thailand, half who had applying bentonite to their fields and half who had not. The average output for those using the clay addition was 18% higher than for non-clay users. Using the clay had enabled some farmers to switch to growing vegetables, which need more fertile soil. This helped to increase their income. The researchers estimated that 200 farmers in northeast Thailand and 400 in Cambodia had adopted the use of clays, and that a further 20,000 farmers were introduced to the new technique.

Land Degradation

Land degradation is a process in which the value of the biophysical environment is affected by a combination of human-induced processes acting upon the land. It is viewed as any change or disturbance to the land perceived to be deleterious or undesirable. Natural hazards are excluded as a cause; however human activities can indirectly affect phenomena such as floods and bush fires.

Serious land degradation in Nauru after the depletion of the phosphate cover through mining

This is considered to be an important topic of the 21st century due to the implications land degradation has upon agronomic productivity, the environment, and its effects on food security. It is estimated that up to 40% of the world's agricultural land is seriously degraded.

Measures

Land degradation is a broad term that can be applied differently across a wide range of scenarios. There are four main ways of looking at land degradation and its impact on the environment around it:

1. A temporary or permanent decline in the productive capacity of the land. This can be seen through a loss of biomass, a loss of actual productivity or in potential productivity, or a loss or change in vegetative cover and soil nutrients.

2. Action in the land's capacity to provide resources for human livelihoods. This can be measured from a base line of past land use.

3. Loss of biodiversity: A loss of range of species or ecosystem complexity as a decline in the environmental quality.

4. Shifting ecological risk: increased vulnerability of the environment or people to destruction or crisis. This is measured through a base line in the form of pre-existing risk of crisis or destruction.

A problem with defining land degradation is that what one group of people might view as degradation, others might view as a benefit or opportunity. For example, planting crops at a location with heavy rainfall and steep slopes would create scientific and environmental concern regarding the risk of soil erosion by water, yet farmers could view the location as a favourable one for high crop yields.

Different Types

Potato field with soil erosion

In addition to the usual types of land degradation that have been known for centuries (water, wind and mechanical erosion, physical, chemical and biological degradation), four other types have emerged in the last 50 years:

- pollution, often chemical, due to agricultural, industrial, mining or commercial activities;
- loss of arable land due to urban construction;
- artificial radioactivity, sometimes accidental;
- land-use constraints associated with armed conflicts.

Overall, 36 types of land degradation can be assessed. All are induced or aggravated by human activities, e.g. sheet erosion, silting, aridification, salinization, urbanization, etc.

Causes

Overgrazing by livestock can lead to land degradation

Land degradation is a global problem largely related to agricultural use. Causes include:

- Land clearance, such as clearcutting and deforestation
- Agricultural depletion of soil nutrients through poor farming practices
- Livestock including overgrazing and overdrafting
- Inappropriate irrigation and overdrafting
- Urban sprawl and commercial development
- Vehicle off-roading
- Quarrying of stone, sand, ore and minerals
- Increase in field size due to economies of scale, reducing shelter for wildlife, as hedgerows and copses disappear
- Exposure of naked soil after harvesting by heavy equipment
- Monoculture, destabilizing the local ecosystem
- Dumping of non-biodegradable trash, such as plastics
- Soil degradation, e.g.
    - Soil contamination
    - Soil erosion
    - Soil acidification
    - Loss of soil carbon

Effects

Overcutting of vegetation occurs when people cut forests, woodlands and shrublands—to obtain timber, fuelwood and other products—at a pace exceeding the rate of natural regrowth. This is frequent in semi-arid environments, where fuelwood shortages are often severe.

Overgrazing is the grazing of natural pastures at stocking intensities above the livestock carrying capacity; the resulting decrease in the vegetation cover is a leading cause of wind and water erosion. It is a significant factor in Afghanistan. The growing population pressure, during 1980-1990, has led to decreases in the already small areas of agricultural land per person in six out of eight countries (14% for India and 22% for Pakistan).

Population pressure also operates through other mechanisms. Improper agricultural practices, for instance, occur only under constraints such as the saturation of good lands under population pressure which leads settlers to cultivate too shallow or too steep soils, plough fallow land before it has recovered its fertility, or attempt to obtain multiple crops by irrigating unsuitable soils.

Soil erosion in a wheat field near Pullman, USA.

High population density is not always related to land degradation. Rather, it is the practices of the human population that can cause a landscape to become degraded. Populations can be a benefit to the land and make it more productive than it is in its natural state. Land degradation is an important factor of internal displacement in many African and Asian countries.

Severe land degradation affects a significant portion of the Earth's arable lands, decreasing the wealth and economic development of nations. As the land resource base becomes less productive, food security is compromised and competition for dwindling resources increases, the seeds of famine and potential conflict are sown.

Sensitivity and Resilience

Sensitivity and resilience are measures of the vulnerability of a landscape to degradation. These two factors combine to explain the degree of vulnerability. Sensitivity is the degree to which a land system undergoes change due to natural forces, human intervention or a combination of both. Resilience is the ability of a landscape to absorb change, without significantly altering the relationship between the relative importance and numbers of individuals and species that compose the community. It also refers to the ability of the region to return to its original state after being changed in some way. The resilience of a landscape can be increased or decreased through human interaction based upon different methods of land-use management. Land that is degraded becomes less resilient than undegraded land, which can lead to even further degradation through shocks to the landscape.

Climate Change

Significant land degradation from seawater inundation, particularly in river deltas and on low-lying islands, is a potential hazard that was identified in a 2007 IPCC report.

IPCC
Assessment reports:
First (1990)
1992 sup.
Second (1995)
Third (2001)
Fourth (2007)
Fifth (2014)
Sixth (2022)
UNFCCC | WMO | UNEP

As a result of sea-level rise from climate change, salinity levels can reach levels where agriculture becomes impossible in very low-lying areas.

Pressure on Ground

Total Stress

When a load is applied to soil, it is carried by the solid grains and the water in the pores. The total vertical stress acting at a point below the ground surface is due to the weight of everything that lies above, including soil, water, and surface loading. Total stress thus increases with depth and with unit weight.

Vertical total stress at depth z, $\sigma_v = \gamma.Z$

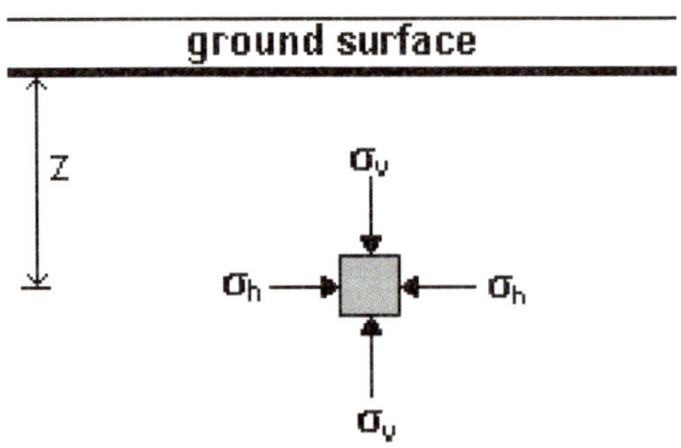

Below a water body, the total stress is the sum of the weight of the soil up to the surface and the weight of water above this. $\sigma_v = \gamma.Z + \gamma_w.Z_w$

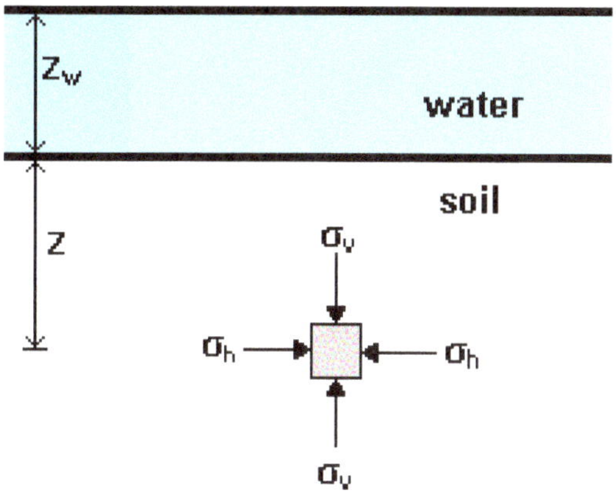

The total stress may also be denoted by σ_z or just σ. It varies with changes in water level and with excavation.

Shear Strength (Soil)

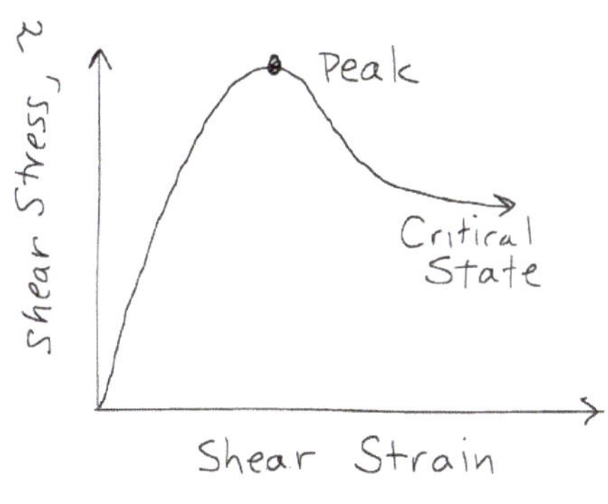

Typical stress strain curve for a drained dilatant soil

Shear strength is a term used in soil mechanics to describe the magnitude of the shear stress that a soil can sustain. The shear resistance of soil is a result of friction and interlocking of particles, and possibly cementation or bonding at particle contacts. Due to interlocking, particulate material may expand or contract in volume as it is subject to shear strains. If soil expands its volume, the density of particles will decrease and the strength will decrease; in this case, the peak strength would be followed by a reduction of shear stress. The stress-strain relationship levels off when the material stops expanding or contracting, and when interparticle bonds are broken. The theoretical state at which the shear stress and density remain constant while the shear strain increases may be called the critical state, steady state, or residual strength.

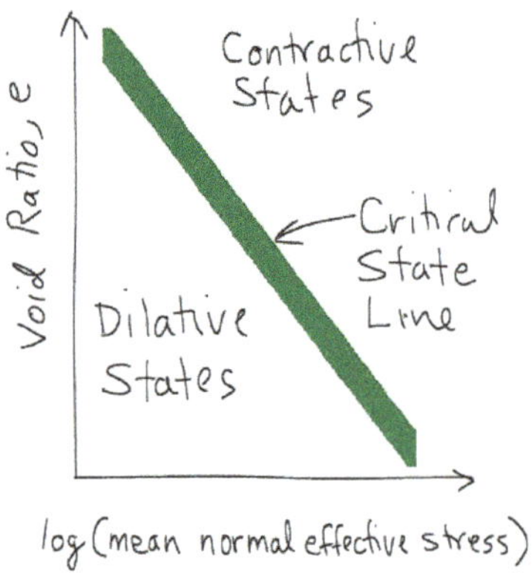

A critical state line separates the dilatant and contractive states for soil

The volume change behavior and interparticle friction depend on the density of the particles, the intergranular contact forces, and to a somewhat lesser extent, other factors such as the rate of shearing and the direction of the shear stress. The average normal intergranular contact force per unit area is called the effective stress.

If water is not allowed to flow in or out of the soil, the stress path is called an *undrained stress path*. During undrained shear, if the particles are surrounded by a nearly incompressible fluid such as water, then the density of the particles cannot change without drainage, but the water pressure and effective stress will change. On the other hand, if the fluids are allowed to freely drain out of the pores, then the pore pressures will remain constant and the test path is called a *drained stress path*. The soil is free to dilate or contract during shear if the soil is drained. In reality, soil is partially drained, somewhere between the perfectly undrained and drained idealized conditions.

The shear strength of soil depends on the effective stress, the drainage conditions, the density of the particles, the rate of strain, and the direction of the strain.

For undrained, constant volume shearing, the Tresca theory may be used to predict the shear strength, but for drained conditions, the Mohr–Coulomb theory may be used.

Two important theories of soil shear are the critical state theory and the steady state theory. There are key differences between the critical state condition and the steady state condition and the resulting theory corresponding to each of these conditions.

Factors Controlling Shear Strength of Soils

The stress-strain relationship of soils, and therefore the shearing strength, is affected (Poulos 1989) by:

1. soil composition (basic soil material): mineralogy, grain size and grain size distribution, shape of particles, pore fluid type and content, ions on grain and in pore fluid.

2. state (initial): Defined by the initial void ratio, effective normal stress and shear stress (stress history). State can be described by terms such as: loose, dense, overconsolidated, normally consolidated, stiff, soft, contractive, dilative, etc.

3. structure: Refers to the arrangement of particles within the soil mass; the manner the particles are packed or distributed. Features such as layers, joints, fissures, slickensides, voids, pockets, cementation, etc., are part of the structure. Structure of soils is described by terms such as: undisturbed, disturbed, remolded, compacted, cemented; flocculent, honey-combed, single-grained; flocculated, deflocculated; stratified, layered, laminated; isotropic and anisotropic.

4. Loading conditions: Effective stress path, i.e., drained, and undrained; and type of loading, i.e., magnitude, rate (static, dynamic), and time history (monotonic, cyclic).

Undrained Strength

This term describes a type of shear strength in soil mechanics as distinct from drained strength.

Conceptually, there is no such thing as *the* undrained strength of a soil. It depends on a number of factors, the main ones being:

- Orientation of stresses
- Stress path
- Rate of shearing
- Volume of material (like for fissured clays or rock mass)

Undrained strength is typically defined by Tresca theory, based on Mohr's circle as:

$$\sigma_1 - \sigma_3 = 2\,S_u$$

Where:

σ_1 is the major principal stress

σ_3 is the minor principal stress

τ is the shear strength $(\sigma_1 - \sigma_3)/2$

hence, $\tau = S_u$ (or sometimes c_u), the undrained strength.

It is commonly adopted in limit equilibrium analyses where the rate of loading is very much greater than the rate at which pore water pressures, that are generated due to the action of shearing the soil, may dissipate. An example of this is rapid loading of sands during an earthquake, or the failure of a clay slope during heavy rain, and applies to most failures that occur during construction.

As an implication of undrained condition, no elastic volumetric strains occur, and thus Poisson's ratio is assumed to remain 0.5 throughout shearing. The Tresca soil model also assumes no plastic volumetric strains occur. This is of significance in more advanced analyses such as in finite element analysis. In these advanced analysis methods, soil models other than Tresca may be used to model the undrained condition including Mohr-Coulomb and critical state soil models such as the modified Cam-clay model, provided Poisson's ratio is maintained at 0.5.

One relationship used extensively by practicing engineers is the empirical observation that the ratio of the undrained shear strength c to the original consolidation stress p' is approximately a constant for a given Over Consolidation Ratio (OCR). This relationship was first formalized by (Henkel 1960) and (Henkel & Wade 1966) who also extended it to show that stress-strain characteristics of remolded clays could also be normalized with respect to the original consolidation stress. The constant c/p relationship can also be derived from theory for both critical-state and steady-state soil mechanics (Joseph 2012). This fundamental, normalization property of the stress-strain curves is found in many clays, and was refined into the empirical SHANSEP (stress history and normalized soil engineering properties) method.(Ladd & Foott 1974).

Drained Shear Strength

The drained shear strength is the shear strength of the soil when pore fluid pressures, generated during the course of shearing the soil, are able to dissipate during shearing. It also applies where no pore water exists in the soil (the soil is dry) and hence pore fluid pressures are negligible. It is commonly approximated using the Mohr-Coulomb equation. (It was called "Coulomb's equation" by Karl von Terzaghi in 1942.) (Terzaghi 1942) combined it with the principle of effective stress.

In terms of effective stresses, the shear strength is often approximated by:

$$\tau = \sigma' \tan(\varphi') + c'$$

Where $\sigma' = (\sigma - u)$, is defined as the effective stress. σ is the total stress applied normal to the shear plane, and u is the pore water pressure acting on the same plane.

φ' = the effective stress friction angle, or the 'angle of internal friction' after Coulomb friction. The coefficient of friction μ is equal to $\tan(\varphi')$. Different values of friction angle can be defined, including the peak friction angle, $\varphi'p$, the critical state friction angle, $\varphi'cv$, or residual friction angle, $\varphi'r$.

c' = is called cohesion, however, it usually arises as a consequence of forcing a straight line to fit through measured values of (τ,σ') even though the data actually falls on a curve. The intercept of the straight line on the shear stress axis is called the cohesion. It is well known that the resulting intercept depends on the range of stresses considered: it is not a fundamental soil property. The curvature (nonlinearity) of the failure envelope occurs because the dilatancy of closely packed soil particles depends on confining pressure.

Critical State Theory

A more advanced understanding of the behaviour of soil undergoing shearing lead to the development of the critical state theory of soil mechanics (Roscoe, Schofield & Wroth 1958). In critical state soil mechanics, a distinct shear strength is identified where the soil undergoing shear does so at a constant volume, also called the 'critical state'. Thus there are three commonly identified shear strengths for a soil undergoing shear:

- Peak strength τ p
- Critical state or constant volume strength τ cv
- Residual strength τ r

The peak strength may occur before or at critical state, depending on the initial state of the soil particles being sheared:

- A loose soil will contract in volume on shearing, and may not develop any peak strength above critical state. In this case 'peak' strength will coincide with the critical state shear strength, once the soil has ceased contracting in volume. It may be stated that such soils do not exhibit a distinct 'peak strength'.
- A dense soil may contract slightly before granular interlock prevents further contraction (granular interlock is dependent on the shape of the grains and their initial packing arrangement). In order to continue shearing once granular interlock has occurred, the soil must dilate (expand in volume). As additional shear force is required to dilate the soil, a 'peak' strength occurs. Once this peak strength caused by dilation has been overcome through continued shearing, the resistance provided by the soil to the applied shear stress reduces (termed "strain softening"). Strain softening will continue until no further changes in volume of the soil occur on continued shearing. Peak strengths are also observed in overconsolidated clays where the natural fabric of the soil must be destroyed prior to reaching constant volume shearing. Other effects that result in peak strengths include cementation and bonding of particles.

The constant volume (or critical state) shear strength is said to be extrinsic to the soil, and independent of the initial density or packing arrangement of the soil grains. In this state the grains being sheared are said to be 'tumbling' over one another, with no significant granular interlock or sliding plane development affecting the resistance to shearing. At this point, no inherited fabric or bonding of the soil grains affects the soil strength.

The residual strength occurs for some soils where the shape of the particles that make up the soil become aligned during shearing (forming a slickenside), resulting in reduced resistance to continued shearing (further strain softening). This is particularly true for most clays that comprise plate-like minerals, but is also observed in some granular soils with more elongate shaped grains. Clays that do not have plate-like minerals (like allophanic clays) do not tend to exhibit residual strengths.

Use in practice: If one is to adopt critical state theory and take c' = 0; τ p may be used, provided the level of anticipated strains are taken into account, and the effects of potential rupture or strain softening to critical state strengths are considered. For large strain deformation, the potential to form slickensided surface with a φ'r should be considered (such as pile driving).

The Critical State occurs at the quasi-static strain rate. It does not allow for differences in shear strength based on different strain rates. Also at the critical state, there is no particle alignment or specific soil structure.

Almost as soon as it was first introduced, the critical state concept has been subject to much criticism--chiefly its inability to match readily available test data from testing a wide variety of soils. This is primarily due to the theories inability to account for particle structure. A major consequence of this is its inability to model strain-softening post peak commonly observed in contractive soils that have anisotropic grain shapes/ properties. Further, an assumption commonly made to make the model mathematically tractable is that shear stress cannot cause volumetric strain nor volumetric stress cause shear strain. Since this is not the case in reality, it is an additional cause of the poor matches to readily available empirical test data. Additionally, critical state elas-to-plastic models assume that elastic strains drives volumetric changes. Since this too is not the case in real soils, this assumption results in poor fits to volume and pore pressure change data.

Steady State (Dynamical Systems Based Soil Shear)

A refinement of the critical state concept is the steady state concept.

The steady state strength is defined as the shear strength of the soil when it is at the steady state condition. The steady state condition is defined (Poulos 1981) as "that state in which the mass is continuously deforming at constant volume, constant normal effective stress, constant shear stress, and constant velocity." Steve J. Poulos, then an Associate Professor of the Soil Mechanics Department of Harvard University, built off a hypothesis that Arthur Casagrande was formulating towards the end of his career. (Poulos 1981) Steady state based soil mechanics is sometimes called "Harvard soil mechanics". The steady state condition is not the same as the "critical state" condition.

The steady state occurs only after all particle breakage if any is complete and all the particles are oriented in a statistically steady state condition and so that the shear stress needed to continue deformation at a constant velocity of deformation does not change. It applies to both the drained and the undrained case.

The steady state has a slightly different value depending on the strain rate at which it is measured. Thus the steady state shear strength at the quasi-static strain rate (the strain rate at which the critical state is defined to occur at) would seem to correspond to the critical state shear strength. However, there is an additional difference between the two states. This is that at the steady state condition the grains position themselves in the steady state structure, whereas no such structure occurs for the critical state. In the case of shearing to large strains for soils with elongated particles, this steady state structure is one where the grains are oriented (perhaps even aligned) in the direction of shear. In the case where the particles are strongly aligned in the direction of shear, the steady state corresponds to the "residual condition."

Three common misconceptions regarding the steady state are that a) it is the same as the critical state (it is not), b) that it applies only to the undrained case (it applies to all forms of drainage), and c) that it does not apply to sands (it applies to any granular material). A primer on the Steady State theory can be found in a report by Poulos (Poulos 1971). Its use in earthquake engineering is described in detail in another publication by Poulos (Poulos 1989).

The difference between the steady state and the critical state is not merely one of semantics as is sometimes thought, and it is incorrect to use the two terms/concepts interchangeably. The additional requirements of the strict definition of the steady state over and above the critical state viz. a constant deformation velocity and statistically constant structure (the steady state structure), places the steady state condition within the framework of dynamical systems theory. This strict definition of the steady state was used to describe soil shear as a dynamical system (Joseph 2009). Dynamical systems are ubiquitous in nature (the Great Red Spot on Jupiter is one example) and mathematicians have extensively studied such systems. The underlying basis of the soil shear dynamical system is simple friction (Joseph 2012).

Mohr–Coulomb Theory

Mohr–Coulomb theory is a mathematical model describing the response of brittle materials such as concrete, or rubble piles, to shear stress as well as normal stress. Most of the classical engineering materials somehow follow this rule in at least a portion of their shear failure envelope. Generally the theory applies to materials for which the compressive strength far exceeds the tensile strength.

In geotechnical engineering it is used to define shear strength of soils and rocks at different effective stresses.

In structural engineering it is used to determine failure load as well as the angle of fracture of a displacement fracture in concrete and similar materials. Coulomb's friction hypothesis is used to determine the combination of shear and normal stress that will cause a fracture of the material. Mohr's circle is used to determine which principal stresses that will produce this combination of shear and normal stress, and the angle of the plane in which this will occur. According to the principle of normality the stress introduced at failure will be perpendicular to the line describing the fracture condition.

It can be shown that a material failing according to Coulomb's friction hypothesis will show the displacement introduced at failure forming an angle to the line of fracture equal to the angle of friction. This makes the strength of the material determinable by comparing the external mechanical work introduced by the displacement and the external load with the internal mechanical work introduced by the strain and stress at the line of failure. By conservation of energy the sum of these must be zero and this will make it possible to calculate the failure load of the construction.

A common improvement of this model is to combine Coulomb's friction hypothesis with Rankine's principal stress hypothesis to describe a separation fracture.

History of the Development

The Mohr–Coulomb theory is named in honour of Charles-Augustin de Coulomb and Christian Otto Mohr. Coulomb's contribution was a 1773 essay entitled "*Essai sur une application des règles des maximis et minimis à quelques problèmes de statique relatifs à l'architecture*" . Mohr developed a generalised form of the theory around the end of the 19th century. As the generalised form affected the interpretation of the criterion, but not the substance of it, some texts continue to refer to the criterion as simply the 'Coulomb criterion'.

Mohr–Coulomb Failure Criterion

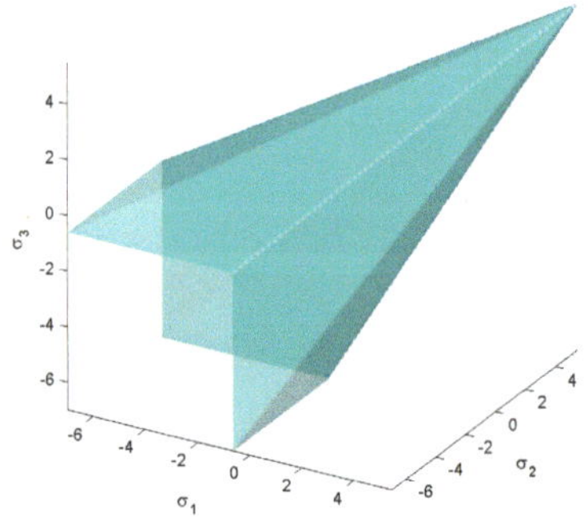

View of Mohr–Coulomb failure surface in 3D space of principal stresses for

The Mohr–Coulomb failure criterion represents the linear envelope that is obtained from a plot of the shear strength of a material versus the applied normal stress. This relation is expressed as

$$\tau = \sigma \tan(\phi) + c$$

where τ is the shear strength, σ is the normal stress, c is the intercept of the failure envelope with the τ axis, and ϕ is the slope of the failure envelope. The quantity c is often called the cohesion and the angle ϕ is called the angle of internal friction . Compression is assumed to be positive in the following discussion. If compression is assumed to be negative then σ should be replaced with $-\sigma$.

If $\phi = 0$, the Mohr–Coulomb criterion reduces to the Tresca criterion. On the other hand, if $\phi = 90^\circ$ the Mohr–Coulomb model is equivalent to the Rankine model. Higher values of ϕ are not allowed.

From Mohr's circle we have

$$\sigma = \sigma_m - \tau_m \sin\phi \; ; \; \tau = \tau_m \cos\phi$$

where

$$\tau_m = \frac{\sigma_1 - \sigma_3}{2} \; ; \; \sigma_m = \frac{\sigma_1 + \sigma_3}{2}$$

and σ_1 is the maximum principal stress and σ_3 is the minimum principal stress.

Therefore, the Mohr–Coulomb criterion may also be expressed as

$$\tau_m = \sigma_m \sin\phi + c\cos\phi \; .$$

This form of the Mohr–Coulomb criterion is applicable to failure on a plane that is parallel to the σ_2 direction.

Mohr–Coulomb Failure Criterion in Three Dimensions

The Mohr–Coulomb criterion in three dimensions is often expressed as

$$\begin{cases} \pm\dfrac{\sigma_1 - \sigma_2}{2} = \left[\dfrac{\sigma_1 + \sigma_2}{2}\right]\sin(\phi) + c\cos(\phi) \\ \pm\dfrac{\sigma_2 - \sigma_3}{2} = \left[\dfrac{\sigma_2 + \sigma_3}{2}\right]\sin(\phi) + c\cos(\phi) \\ \pm\dfrac{\sigma_3 - \sigma_1}{2} = \left[\dfrac{\sigma_3 + \sigma_1}{2}\right]\sin(\phi) + c\cos(\phi) \end{cases}$$

The Mohr–Coulomb failure surface is a cone with a hexagonal cross section in deviatoric stress space.

The expressions for τ and σ can be generalized to three dimensions by developing expressions for the normal stress and the resolved shear stress on a plane of arbitrary orientation with respect to the coordinate axes (basis vectors). If the unit normal to the plane of interest is

$$\mathbf{n} = n_1\,\mathbf{e}_1 + n_2\,\mathbf{e}_2 + n_3\,\mathbf{e}_3$$

where $\mathbf{e}_i,\ \ i = 1,2,3$ are three orthonormal unit basis vectors, and if the principal stresses $\sigma_1, \sigma_2, \sigma_3$ are aligned with the basis vectors $\mathbf{e}_1, \mathbf{e}_2, \mathbf{e}_3$, then the expressions for σ, τ are

$$\sigma = n_1^2\sigma_1 + n_2^2\sigma_2 + n_3^2\sigma_3$$

$$\tau = \sqrt{(n_1\sigma_1)^2 + (n_2\sigma_2)^2 + (n_3\sigma_3)^2 - \sigma^2}$$

$$= \sqrt{n_1^2 n_2^2(\sigma_1 - \sigma_2)^2 + n_2^2 n_3^2(\sigma_2 - \sigma_3)^2 + n_3^2 n_1^2(\sigma_3 - \sigma_1)^2}.$$

The Mohr–Coulomb failure criterion can then be evaluated using the usual expression

$$\tau = \sigma\ \tan(\phi) + c$$

for the six planes of maximum shear stress.

Derivation of Normal and Shear Stress on a Plane

Let the unit normal to the plane of interest be

$$\mathbf{n} = n_1\,\mathbf{e}_1 + n_2\,\mathbf{e}_2 + n_3\,\mathbf{e}_3$$

where $\mathbf{e}_i,\ \ i = 1,2,3$ are three orthonormal unit basis vectors. Then the traction vector on the plane is given by

$$\mathbf{t} = n_i\,\sigma_{ij}\,\mathbf{e}_j \quad \text{(repeated indices indicate summation)}$$

The magnitude of the traction vector is given by

$$|\,\mathbf{t}\,| = \sqrt{(n_j\,\sigma_{1j})^2 + (n_k\,\sigma_{2k})^2 + (n_l\,\sigma_{3l})^2} \quad \text{(repeated indices indicate summation)}$$

Then the magnitude of the stress normal to the plane is given by

$$\sigma = \mathbf{t}\cdot\mathbf{n} = n_i\,\sigma_{ij}\,n_j \quad \text{(repeated indices indicate summation)}$$

The magnitude of the resolved shear stress on the plane is given by

$$\tau = \sqrt{|\,\mathbf{t}\,|^2 - \sigma^2}$$

In terms of components, we have

$$\sigma = n_1^2\sigma_{11} + n_2^2\sigma_{22} + n_3^2\sigma_{33} + 2(n_1n_2\sigma_{12} + n_2n_3\sigma_{23} + n_3n_1\sigma_{31})$$

$$\tau = \sqrt{(n_1\sigma_{11} + n_2\sigma_{12} + n_3\sigma_{31})^2 + (n_1\sigma_{12} + n_2\sigma_{22} + n_3\sigma_{23})^2 + (n_1\sigma_{31} + n_2\sigma_{23} + n_3\sigma_{33})^2 - \sigma^2}$$

If the principal stresses $\sigma_1, \sigma_2, \sigma_3$ are aligned with the basis vectors $\mathbf{e}_1, \mathbf{e}_2, \mathbf{e}_3$, then the expressions for σ, τ are

$$\sigma = n_1^2\sigma_1 + n_2^2\sigma_2 + n_3^2\sigma_3$$

$$\tau = \sqrt{(n_1\sigma_1)^2 + (n_2\sigma_2)^2 + (n_3\sigma_3)^2 - \sigma^2}$$

$$= \sqrt{n_1^2n_2^2(\sigma_1 - \sigma_2)^2 + n_2^2n_3^2(\sigma_2 - \sigma_3)^2 + n_3^2n_1^2(\sigma_3 - \sigma_1)^2}$$

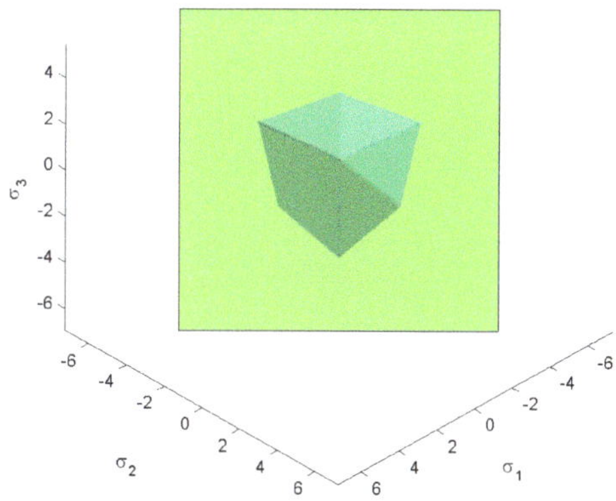

Mohr–Coulomb yield surface in the -plane for $c = 2, \phi = 20^\circ$

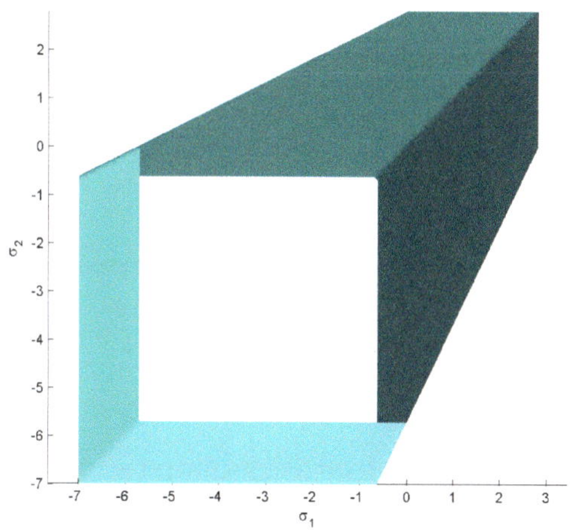

Trace of the Mohr–Coulomb yield surface in the $\sigma_1 - \sigma_2$ -plane for $c = 2, \phi = 20^\circ$

Mohr–Coulomb Failure Surface in Haigh–Westergaard Space

The Mohr–Coulomb failure (yield) surface is often expressed in Haigh–Westergaad coordinates. For example, the function

$$\frac{\sigma_1-\sigma_3}{2}=\frac{\sigma_1+\sigma_3}{2}\sin\phi+c\cos\phi$$

can be expressed as

$$\left[\sqrt{3}\,\sin\left(\theta+\frac{\pi}{3}\right)-\sin\phi\cos\left(\theta+\frac{\pi}{3}\right)\right]\rho-\sqrt{2}\sin(\phi)\xi=\sqrt{6}c\cos\phi.$$

Alternatively, in terms of the invariants p,q,r we can write

$$\left[\frac{1}{\sqrt{3}\cos\phi}\sin\left(\theta+\frac{\pi}{3}\right)-\frac{1}{3}\tan\phi\,\cos\left(\theta+\frac{\pi}{3}\right)\right]q-p\,\tan\phi=c$$

where

$$\theta=\frac{1}{3}\arccos\left[\left(\frac{r}{q}\right)^3\right].$$

Derivation of alternative forms of Mohr–Coulomb yield function

We can express the yield function

$$\frac{\sigma_1-\sigma_3}{2}=\frac{\sigma_1+\sigma_3}{2}\sin\phi+c\cos\phi$$

as

$$\sigma_1\frac{(1-\sin\phi)}{2c\,\cos\phi}-\sigma_3\frac{(1+\sin\phi)}{2c\,\cos\phi}=1\,.$$

The Haigh–Westergaard invariants are related to the principal stresses by

$$\sigma_1=\frac{1}{\sqrt{3}}\xi+\sqrt{\frac{2}{3}}\rho\,\cos\theta\;;\;\sigma_3=\frac{1}{\sqrt{3}}\xi+\sqrt{\frac{2}{3}}\rho\,\cos\left(\theta+\frac{2\pi}{3}\right).$$

Plugging into the expression for the Mohr–Coulomb yield function gives us

$$-\sqrt{2}\,\xi\,\sin\phi+\rho[\cos\theta-\cos(\theta+2\pi/3)]-\rho\sin\phi[\cos\theta+\cos(\theta+2\pi/3)]=\sqrt{6}\,c\,\cos\phi$$

Using trigonometric identities for the sum and difference of cosines and rearrangement gives us the expression of the Mohr–Coulomb yield function in terms of ξ, ρ, θ.

We can express the yield function in terms of p, q by using the relations

$$\xi = \sqrt{3}\, p \;;\; \rho = \sqrt{\frac{2}{3}}\, q$$

and straightforward substitution.

Mohr–Coulomb Yield and Plasticity

The Mohr–Coulomb yield surface is often used to model the plastic flow of geomaterials (and other cohesive-frictional materials). Many such materials show dilatational behavior under triaxial states of stress which the Mohr–Coulomb model does not include. Also, since the yield surface has corners, it may be inconvenient to use the original Mohr–Coulomb model to determine the direction of plastic flow (in the flow theory of plasticity).

A common approach is to use a non-associated plastic flow potential that is smooth. An example of such a potential is the function

$$g := \sqrt{(\alpha c_y \tan\psi)^2 + G^2(\phi,\theta) q^2} - p \tan\phi$$

where α is a parameter, c_y is the value of c when the plastic strain is zero (also called the initial cohesion yield stress), ψ is the angle made by the yield surface in the Rendulic plane at high values of p (this angle is also called the dilation angle), and (ϕ,θ) is an appropriate function that is also smooth in the deviatoric stress plane.

Shearing Strength of Granular Soils

For granular soils with $c = 0$,

$$s = \sigma' \tan\phi$$

The determination of friction angle ϕ is commonly accomplished by one of two methods; the direct shear test or the triaxial test. The test procedures are given below.

Direct Shear Test

A schematic diagram of the direct shear test equipment is shown in figure. Basically, the test equipment consists of a metal shear box into which the soil specimen is placed. The specimen can be square or circular in plan, about 3 to 4 (19.35 to 25.80) in area, and about 1 in (25.4 mm) in height. The box is split horizontally into two halves. Normal force on the specimen is applied from the top of the shear box by dead weights. The normal stress on the specimens obtained by the application of dead weights can be as high as . Shear force is applied to the side of the top half of the box to cause failure in the soil specimen. (The two porous stones shown in figure are not required for tests on dry soil). During the test, the shear displacement of the top half of the box and the change in specimen thickness are recorded by the use of horizontal and vertical dial gauges.

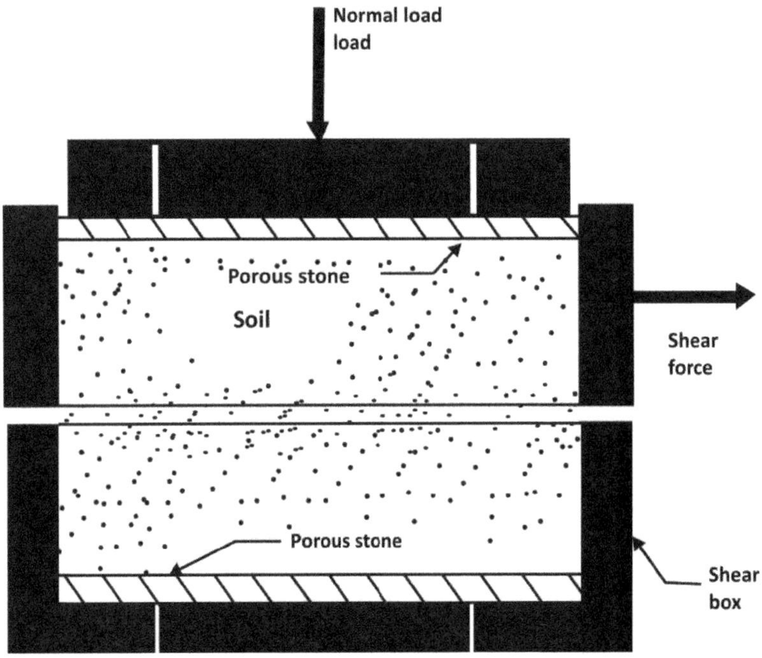

Direct shear test arrangement

Figure shows the nature of the results of typical direct shear tests in loose, medium, and dense sands. The following observations can be drawn

1. In dense and medium sands, shear stress increases with shear displacement to a maximum or peak value τ_m and then decreases to an approximately constant value τ_{cv} at large shear displacement. This constant stress τ_{cv} is the ultimate shear stress.

2. For loose sands, the shear stress increases with shear displacement to a maximum value and then remains constant.

3. For dense and medium sands, the volume of the specimen initially decreases and then increases with shear displacement. At large values of shear displacement, the volume of the specimen remains approximately constant.

4. For loose sands, the volume of the specimen gradually decreases to a certain value and remains approximately constant thereafter.

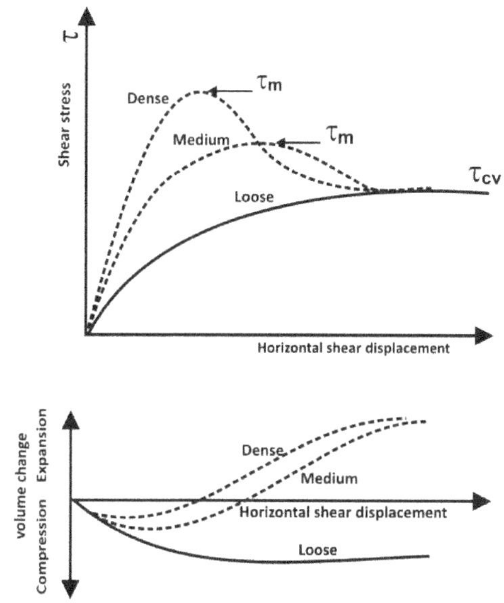

Direct shear test results in loose, medium and dense sands

If dry sand is used for the test, the pore water pressure u is equal to zero, and so the total normal stress σ is equal to the effective stress σ'. The test may be repeated for several normal stresses. The angle of friction ϕ for the sand can be determined by plotting a graph of the maximum or peak shear stresses vs. the corresponding normal stresses, as shown in figure. The Mohr-Coulomb failure envelope can be determined by drawing a straight line through the origin and the points representing the experimental results. The slope of this line will give the peak friction angle ϕ of the soil. Similarly, the ultimate friction angle ϕ_{cv} can be determined by plotting the ultimate shear stresses τ_{cv} vs. The corresponding normal stresses, as shown in figure. The ultimate friction angle ϕ_{cv} represents a condition of shearing at constant volume of the specimen.

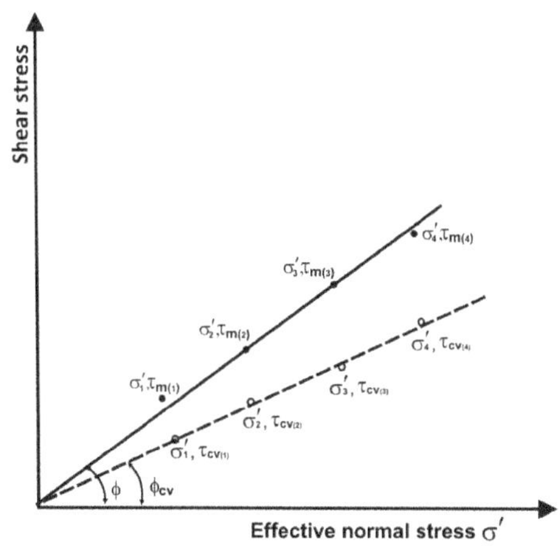

Determination of peak and ultimate friction angle from direct shear test

Some typical values of ϕ *and* ϕ_{cv} for granular soils are given table.

Table Typical values of for granular soils

Type of soil	deg	
Sand: round grains		
Loose	28 to30	
Medium	30 to 35	26 to 30
Dense	35 to 38	
Sand: angular grains		
Loose	30 to 35	
Medium	35 to 40	30 to 35
Dense	40 to 45	
Sandy gravel	34 to 48	33 to 36

Triaxial Test

A schematic diagram of a triaxial test equipment is shown in figure. In this type of test, a soil specimen about 1.5 in (38.1 mm) in diameter and 3 in (76.2 mm) in length is generally used. The specimen is enclosed inside a thin rubber membrane and placed inside a cylindrical plastic chamber. For conducting the test, the chamber is usually filled with water or glycerin. The specimen is subjected to a confining pressure σ_3 by application of pressure to the fluid in the chamber. (Air can sometimes be used as a medium for applying the confining pressure). Connections to measure drainage into or out of the specimen or pressure in the pore water are provided. To cause shear failure in the soil, an axial stress $\Delta\sigma$ is applied through a vertical loading ram. This is also referred to as deviator stress. For determination of ϕ dry or fully saturated soil can be used. If saturated soil is used, the drainage connection is kept open during the application of the confining pressure and the deviator stress. Thus, during the test the excess pore water pressure in the specimen is equal to zero. The volume of the water drained from the specimen during the test provides a measure of the volume change of the specimen.

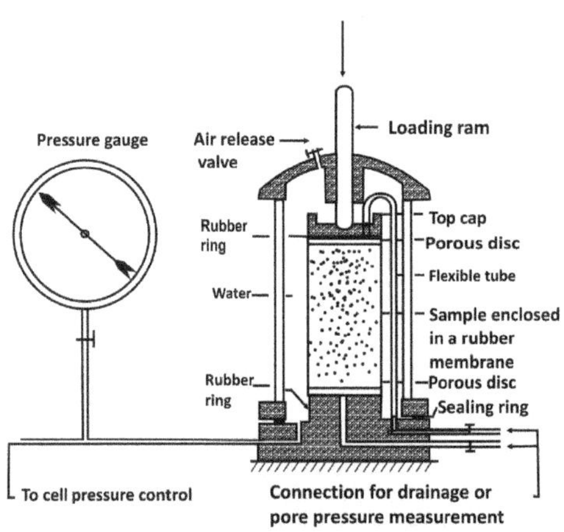

Triaxial test equipment.

For drained test, the total stress is equal to the effective stress. Thus, the major effective principal stress is $\sigma_1' = \sigma_1 = \sigma_3 + \Delta\sigma$; the minor effective principal stress is $\sigma_3' = \sigma_3$; and the intermediate effective principal stress is $\sigma_2' = \sigma_3'$.

At failure, the major effective principal stress is equal to $\sigma_3 + \Delta\sigma_f$ where $\Delta\sigma_f$ is the deviator stress at failure, and the minor effective principal stress is σ_3. Figure shows the nature of the variation of $\Delta\sigma$ with axial strain for loose and dense granular soils. Several tests with similar specimens can be conducted by the using different confining pressure σ_3. The value of the soil peak friction angle ϕ can be determined by plotting effective-stress Mohr's circles for various tests and drawing a common tangent to these Mohr's circles passing through the origin. This is shown in figure a. The angle that this envelope makes with the normal stress axis is equal to ϕ. It can be seen from figure b that

$$\sin\phi = \frac{\overline{ab}}{\overline{oa}} = \frac{(\sigma_1' - \sigma_3')/2}{(\sigma_1' + \sigma_3')/2}$$

$$Or\ \phi = \sin^{-1}\frac{\overline{ab}}{\overline{oa}} = \frac{(\sigma_1' - \sigma_3')}{(\sigma_1' + \sigma_3')}_{failure}$$

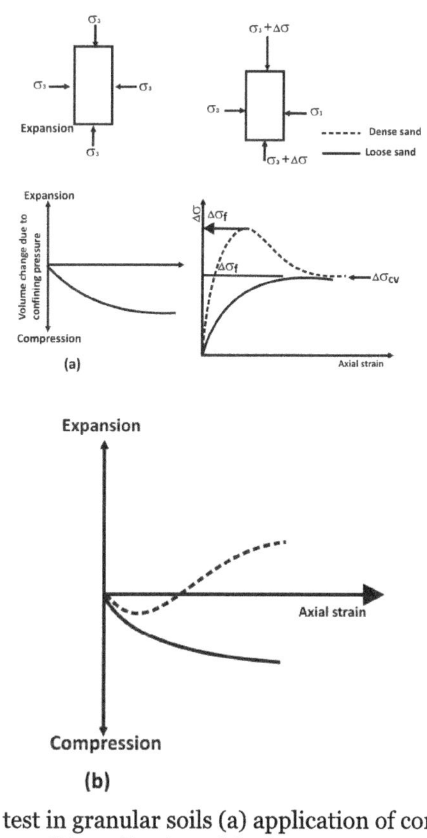

Drained triaxial test in granular soils (a) application of confining pressure (b) application of deviator stress

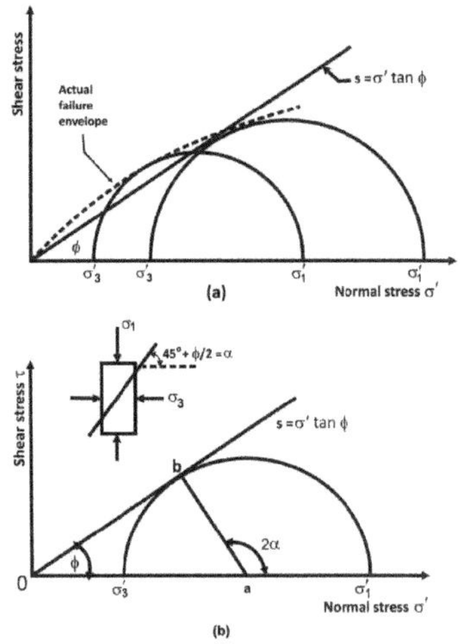

Drained triaxial test results

However, it must be pointed out that in figure a the failure envelope defined by the equation $s = \sigma' \tan\phi$ is an approximation to the actual curved failure envelope. The ultimate friction angle ϕ_{cv} for a given test can be also be determined from the equation

$$\phi_{cv} = \sin^{-1}\left[\frac{\sigma'_{1(cv)} - \sigma'_3}{\sigma'_{1(cv)} + \sigma'_3}\right]$$

Where $\sigma'_{1(cv)} = \sigma'_3 + \Delta\sigma_{(cv)}$ For similar soils, the friction angle ϕ determined by triaxial tests is slightly lower $\left(0\ to\ 3^\circ\right)$ than that obtained from direct shear tests.

The axial compression triaxial test described above is the conventional type. However, the loading process on the specimen in a triaxial chamber can be varied in several ways. In general, the tests can be divided into two major groups: axial compression tests and axial extensionn tests.

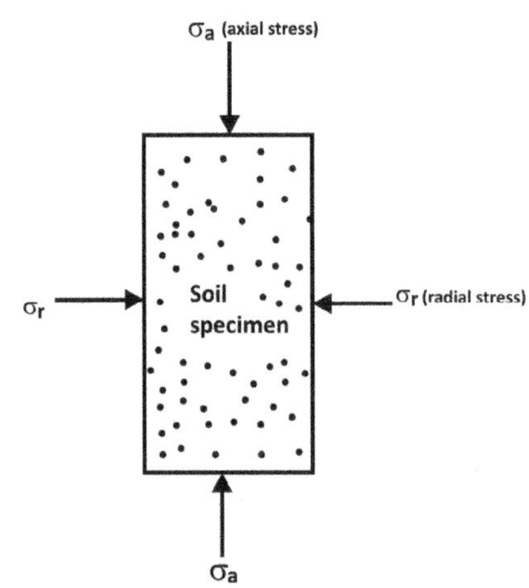

Soil specimen with axial and radial stress

Total Stress Parameters

UU Tests:

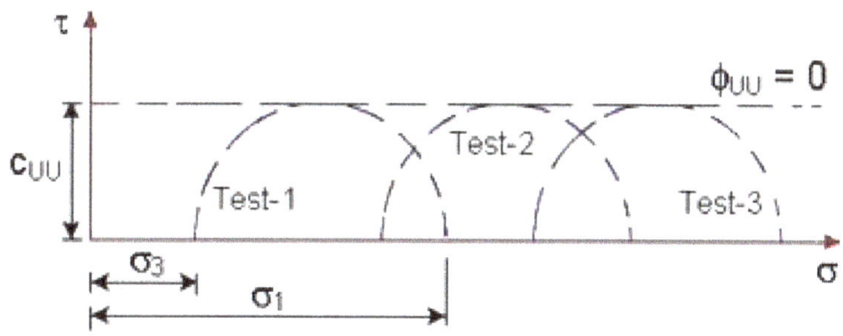

All Mohr circles for UU test plotted in terms of total stresses have the same diameter.

The failure envelope is a horizontal straight line and hence $\phi_{UU} = 0$

It can be represented by the equation:

$$\tau_j = c_{UU} = \frac{\sigma_1 - \sigma_3}{2}$$

CU & CD Tests:

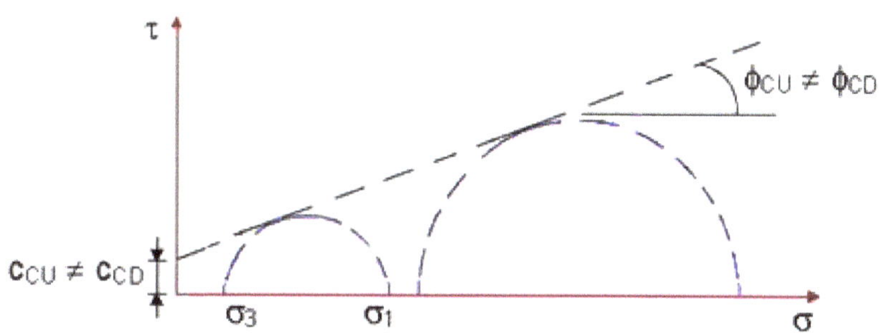

For tests involving drainage in the first stage, when Mohr circles are plotted in terms of total stresses, the diameter increases with the confining pressure. The resulting failure envelope is an inclined line with an intercept on the vertical axis.

It is also observed that $c_{CU} \neq c_{CD}$ and $\phi_{CU} \neq \phi_{CD}$

It can be stated that for identical soil samples tested under different triaxial conditions of UU, CU and CD tests, the failure envelope is not unique.

References

- Johnson, D.L., S.H. Ambrose, T.J. Bassett, M.L. Bowen, D.E. Crummey, J.S. Isaacson, D.N. Johnson, P. Lamb, M. Saul, and A.E. Winter-Nelson. 1997. Meanings of environmental terms. Journal of Environmental Quality 26: 581-589

- Juvinal, Robert C. & Marshek, Kurt .; Fundamentals of machine component design. – 2nd ed., 1991, pp. 217, ISBN 0-471-62281-8
- Ian Sample (2007-08-31). "Global food crisis looms as climate change and population growth strip fertile land". The Guardian. Retrieved 2008-07-23
- Henkel, D. J.; Wade, N. H. (1966), "Plane Strain Tests on a Saturated Remolded Clay", Journal of the Soil Mechanics and Foundation Division, ASCE, 92 (SM 6), pp. 67–80
- Joseph, P. G. (2012), "Physical Basis and Validation of a Constitutive Model for Soil Shear Derived from Micro-Structural Changes" (PDF), International Journal of Geomechanics, doi:10.1061/(ASCE)GM.1943-5622.0000209
- Conacher, Arthur; Conacher, Jeanette (1995). Rural Land Degradation in Australia. South Melbourne, Victoria: Oxford University Press Australia. p. 2. ISBN 0-19-553436-0
- Ladd, C. C.; Foott, R. (1974), "New Design Procedure for Stability of Soft Clays", Journal of Geotechnical Engineering, 100 (GT7), pp. 763–786
- Joseph, P. G. (2017), Dynamical Systems-Based Soil Mechanics, Leiden: CRC Press/Balkema, p. 138, ISBN 9781138723221

Soil Contamination and Soil Permeability

Soil contains pores which have gases and liquids present in them. These pores also help in passage of water. Permeability depends on this passage as its value can be determined on the size of pores. Other factors that affect permeability are particle shape, distribution of particle sizes, and soil structure. Also discussed are the concepts of soil contamination, mining space, etc. Soil contamination and permeability are one of the significant and important topics related to soil mechanics. The following chapter unfolds its crucial aspects in a critical yet systematic manner.

Soil Contamination

Excavation showing soil contamination at a disused gasworks.

Soil contamination or soil pollution as part of land degradation is caused by the presence of xenobiotic (human-made) chemicals or other alteration in the natural soil environment. It is typically caused by industrial activity, agricultural chemicals, or improper disposal of waste. The most common chemicals involved are petroleum hydrocarbons, polynuclear aromatic hydrocarbons (such as naphthalene and benzo(a)pyrene), solvents, pesticides, lead, and other heavy metals. Contamination is correlated with the degree of industrialization and intensity of chemical usage.

The concern over soil contamination stems primarily from health risks, from direct contact with the contaminated soil, vapors from the contaminants, and from secondary contamination of water supplies within and underlying the soil. Mapping of contaminated soil sites and the resulting cleanups are time consuming and expensive tasks, requiring extensive amounts of geology, hydrology, chemistry, computer modeling skills, and GIS in Environmental Contamination, as well as an appreciation of the history of industrial chemistry.

In North America and Western Europe the extent of contaminated land is best known, with many of countries in these areas having a legal framework to identify and deal with this environmental problem. Developing countries tend to be less tightly regulated despite some of them having undergone significant industrialization.

Causes

Soil pollution can be caused by the following

- Oil drilling
- Mining and activities by other heavy industries
- Accidental spills as may happen during activities, etc.
- Corrosion of underground storage tanks (including piping used to transmit the contents)
- Acid rain (in turn caused by air pollution)
- Intensive farming
- Agrochemicals, such as pesticides, herbicides and fertilizers
- Industrial accidents
- Road debris
- Drainage of contaminated surface water into the soil
- Waste disposal
    - Oil and fuel dumping
    - Nuclear wastes
    - Direct discharge of industrial wastes to the soil
    - Discharge of sewage sludge
    - Landfill and illegal dumping
    - coal ash
    - Electronic waste
    - ammunitions and agents of war

The most common chemicals involved are petroleum hydrocarbons, solvents, pesticides, lead, and other heavy metals.

In a wider sense, genetically modified plants (GMP) can count as a risk factor for soils, because of their potential to affect the soil fauna. Any activity that leads to other forms of soil degradation (erosion, compaction, etc.) may indirectly worsen the contamination effects in that soil remediation becomes more tedious.

Historical deposition of coal ash used for residential, commercial, and industrial heating, as well as for industrial processes such as ore smelting, were a common source of contamination in areas that were industrialized before about 1960. Coal naturally concentrates lead and zinc during its formation, as well as other heavy metals to a lesser degree. When the coal is burned, most of these metals become concentrated in the ash (the principal exception being mercury). Coal ash and slag may contain sufficient lead to qualify as a "characteristic hazardous waste", defined in the USA as containing more than 5 mg/l of extractable lead using the TCLP procedure. In addition to lead, coal ash typically contains variable but significant concentrations of polynuclear aromatic hydrocarbons (PAHs; e.g., benzo(a)anthracene, benzo(b)fluoranthene, benzo(k)fluoranthene, benzo(a)pyrene, indeno(cd)pyrene, phenanthrene, anthracene, and others). These PAHs are known human carcinogens and the acceptable concentrations of them in soil are typically around 1 mg/kg. Coal ash and slag can be recognised by the presence of off-white grains in soil, gray heterogeneous soil, or (coal slag) bubbly, vesicular pebble-sized grains.

Treated sewage sludge, known in the industry as biosolids, has become controversial as a "fertilizer". As it is the byproduct of sewage treatment, it generally contains more contaminants such as organisms, pesticides, and heavy metals than other soil.

In the European Union, the Urban Waste Water Treatment Directive allows sewage sludge to be sprayed onto land. The volume is expected to double to 185,000 tons of dry solids in 2005. This has good agricultural properties due to the high nitrogen and phosphate content. In 1990/1991, 13% wet weight was sprayed onto 0.13% of the land; however, this is expected to rise 15 fold by 2005. Advocates say there is a need to control this so that pathogenic microorganisms do not get into water courses and to ensure that there is no accumulation of heavy metals in the top soil.

Pesticides and Herbicides

A pesticide is a substance or mixture of substances used to kill a pest. A pesticide may be a chemical substance, biological agent (such as a virus or bacteria), antimicrobial, disinfectant or device used against any pest. Pests include insects, plant pathogens, weeds, mollusks, birds, mammals, fish, nematodes (roundworms) and microbes that compete with humans for food, destroy property, spread or are a vector for disease or cause a nuisance. Although there are benefits to the use of pesticides, there are also drawbacks, such as potential toxicity to humans and other organisms.

Herbicides are used to kill weeds, especially on pavements and railways. They are similar to auxins and most are biodegradable by soil bacteria. However, one group derived from trinitrotoluene (2:4 D and 2:4:5 T) have the impurity dioxin, which is very toxic and causes fatality even in low concentrations. Another herbicide is Paraquat. It is highly toxic but it rapidly degrades in soil due to the action of bacteria and does not kill soil fauna.

Insecticides are used to rid farms of pests which damage crops. The insects damage not only standing crops but also stored ones and in the tropics it is reckoned that one third of the total production is lost during food storage. As with fungicides, the first insecticides used in the nineteenth century were inorganic e.g. Paris Green and other compounds of arsenic. Nicotine has also been used since the late eighteenth century.

There are now two main groups of synthetic insecticides -

1. Organochlorines include DDT, Aldrin, Dieldrin and BHC. They are cheap to produce, potent and persistent. DDT was used on a massive scale from the 1930s, with a peak of 72,000 tonnes used 1970. Then usage fell as the harmful environmental effects were realized. It was found worldwide in fish and birds and was even discovered in the snow in the Antarctic. It is only slightly soluble in water but is very soluble in the bloodstream. It affects the nervous and endocrine systems and causes the eggshells of birds to lack calcium causing them to be easily breakable. It is thought to be responsible for the decline of the numbers of birds of prey like ospreys and peregrine falcons in the 1950s - they are now recovering. As well as increased concentration via the food chain, it is known to enter via permeable membranes, so fish get it through their gills. As it has low water solubility, it tends to stay at the water surface, so organisms that live there are most affected. DDT found in fish that formed part of the human food chain caused concern, but the levels found in the liver, kidney and brain tissues was less than 1 ppm and in fat was 10 ppm, which was below the level likely to cause harm. However, DDT was banned in the UK and the United States to stop the further buildup of it in the food chain. U.S. manufacturers continued to sell DDT to developing countries, who could not afford the expensive replacement chemicals and who did not have such stringent regulations governing the use of pesticides..

2. Organophosphates, e.g. parathion, methyl parathion and about 40 other insecticides are available nationally. Parathion is highly toxic, methyl-parathion is less so and Malathion is generally considered safe as it has low toxicity and is rapidly broken down in the mammalian liver. This group works by preventing normal nerve transmission as cholinesterase is prevented from breaking down the transmitter substance acetylcholine, resulting in uncontrolled muscle movements.

Agents of War

The disposal of munitions, and a lack of care in manufacture of munitions caused by the urgency of production, can contaminate soil for extended periods. There is little published evidence on this type of contamination largely because of restrictions placed by Governments of many countries on the publication of material related to war effort. However, mustard gas stored during World War II has contaminated some sites for up to 50 years and the testing of Anthrax as a potential biological weapon contaminated the whole island of Gruinard

Health Effects

Contaminated or polluted soil directly affects human health through direct contact with soil or via inhalation of soil contaminants which have vaporized; potentially greater threats are posed by the infiltration of soil contamination into groundwater aquifers used for human consumption, sometimes in areas apparently far removed from any apparent source of above ground contamination. This tends to result in the development of pollution-related diseases.

Health consequences from exposure to soil contamination very greatly depending on pollutant type, pathway of attack and vulnerability of the exposed population. Chronic exposure to chromium, lead and other metals, petroleum, solvents, and many pesticide and herbicide formulations can be carcinogenic, can cause congenital disorders, or can cause other chronic health conditions. Industrial or man-made concentrations of naturally occurring substances, such as nitrate and ammonia associated with livestock manure from agricultural operations, have also been identified as health hazards in soil and groundwater.

Chronic exposure to benzene at sufficient concentrations is known to be associated with higher incidence of leukemia. Mercury and cyclodienes are known to induce higher incidences of kidney damage and some irreversible diseases. PCBs and cyclodienes are linked to liver toxicity. Organophosphates and carbonates can induce a chain of responses leading to neuromuscular blockage. Many chlorinated solvents induce liver changes, kidney changes and depression of the central nervous system. There is an entire spectrum of further health effects such as headache, nausea, fatigue, eye irritation and skin rash for the above cited and other chemicals. At sufficient dosages a large number of soil contaminants can cause death by exposure via direct contact, inhalation or ingestion of contaminants in groundwater contaminated through soil.

The Scottish Government has commissioned the Institute of Occupational Medicine to undertake a review of methods to assess risk to human health from contaminated land. The overall aim of the project is to work up guidance that should be useful to Scottish Local Authorities in assessing whether sites represent a significant possibility of significant harm (SPOSH) to human health. It is envisaged that the output of the project will be a short document providing high level guidance on health risk assessment with reference to existing published guidance and methodologies that have been identified as being particularly relevant and helpful. The project will examine how policy guidelines have been developed for determining the acceptability of risks to human health and propose an approach for assessing what constitutes unacceptable risk in line with the criteria for SPOSH as defined in the legislation and the Scottish Statutory Guidance.

Ecosystem Effects

Not unexpectedly, soil contaminants can have significant deleterious consequences for ecosystems. There are radical soil chemistry changes which can arise from the presence of many hazardous chemicals even at low concentration of the contaminant species. These changes can manifest in the alteration of metabolism of endemic microorganisms and arthropods resident in a given soil environment. The result can be virtual eradication of some of the primary food chain, which in turn could have major consequences for predator or consumer species. Even if the chemical effect on lower life forms is small, the lower pyramid levels of the food chain may ingest alien chemicals, which normally become more concentrated for each consuming rung of the food chain. Many of these effects are now well known, such as the concentration of persistent DDT materials for avian consumers, leading to weakening of egg shells, increased chick mortality and potential extinction of species.

Effects occur to agricultural lands which have certain types of soil contamination. Contaminants typically alter plant metabolism, often causing a reduction in crop yields. This has a secondary effect upon soil conservation, since the languishing crops cannot shield the Earth's soil from erosion. Some of these chemical contaminants have long half-lives and in other cases derivative chemicals are formed from decay of primary soil contaminants.

Cleanup Options

Cleanup or environmental remediation is analyzed by environmental scientists who utilize field measurement of soil chemicals and also apply computer models (GIS in Environmental Contamination) for analyzing transport and fate of soil chemicals. Various technologies have been developed for remediation of oil-contaminated soil/ sediments There are several principal strategies for remediation:

- Excavate soil and take it to a disposal site away from ready pathways for human or sensitive ecosystem contact. This technique also applies to dredging of bay muds containing toxins.
- Aeration of soils at the contaminated site (with attendant risk of creating air pollution)
- Thermal remediation by introduction of heat to raise subsurface temperatures sufficiently high to volatize chemical contaminants out of the soil for vapour extraction. Technologies include ISTD, electrical resistance heating (ERH), and ET-DSP^{tm}.

- Bioremediation, involving microbial digestion of certain organic chemicals. Techniques used in bioremediation include landfarming, biostimulation and bioaugmentating soil biota with commercially available microflora.
- Extraction of groundwater or soil vapor with an active electromechanical system, with subsequent stripping of the contaminants from the extract.
- Containment of the soil contaminants (such as by capping or paving over in place).
- Phytoremediation, or using plants (such as willow) to extract heavy metals.
- Mycoremediation, or using fungus to metabolize contaminants and accumulate heavy metals.
- Remediation of oil contaminated sediments with self-collapsing air microbubbles.

By Country

Various national standards for concentrations of particular contaminants include the United States EPA Region 9 Preliminary Remediation Goals (U.S. PRGs), the U.S. EPA Region 3 Risk Based Concentrations (U.S. EPA RBCs) and National Environment Protection Council of Australia Guideline on Investigation Levels in Soil and Groundwater.

People's Republic of China

The immense and sustained growth of the People's Republic of China since the 1970s has exacted a price from the land in increased soil pollution. The State Environmental Protection Administration believes it to be a threat to the environment, to food safety and to sustainable agriculture. According to a scientific sampling, 150 million mu (100,000 square kilometres) of China's cultivated land have been polluted, with contaminated water being used to irrigate a further 32.5 million mu (21,670 square kilometres) and another 2 million mu (1,300 square kilometres) covered or destroyed by solid waste. In total, the area accounts for one-tenth of China's cultivatable land, and is mostly in economically developed areas. An estimated 12 million tonnes of grain are contaminated by heavy metals every year, causing direct losses of 20 billion yuan (US$2.57 billion).

United Kingdom

Generic guidance commonly used in the United Kingdom are the Soil Guideline Values published by the Department for Environment, Food and Rural Affairs (DEFRA) and the Environment Agency. These are screening values that demonstrate the minimal acceptable level of a substance. Above this there can be no assurances in terms of significant risk of harm to human health. These have been derived using the Contaminated

Land Exposure Assessment Model (CLEA UK). Certain input parameters such as Health Criteria Values, age and land use are fed into CLEA UK to obtain a probabilistic output.

Guidance by the Inter Departmental Committee for the Redevelopment of Contaminated Land (ICRCL) has been formally withdrawn by DEFRA, for use as a prescriptive document to determine the potential need for remediation or further assessment.

The CLEA model published by DEFRA and the Environment Agency (EA) in March 2002 sets a framework for the appropriate assessment of risks to human health from contaminated land, as required by Part IIA of the Environmental Protection Act 1990. As part of this framework, generic Soil Guideline Values (SGVs) have currently been derived for ten contaminants to be used as "intervention values". These values should not be considered as remedial targets but values above which further detailed assessment should be considered.

Three sets of CLEA SGVs have been produced for three different land uses, namely

- residential (with and without plant uptake)
- allotments
- commercial/industrial

It is intended that the SGVs replace the former ICRCL values. It should be noted that the CLEA SGVs relate to assessing chronic (long term) risks to human health and do not apply to the protection of ground workers during construction, or other potential receptors such as groundwater, buildings, plants or other ecosystems. The CLEA SGVs are not directly applicable to a site completely covered in hardstanding, as there is no direct exposure route to contaminated soils.

To date, the first ten of fifty-five contaminant SGVs have been published, for the following: arsenic, cadmium, chromium, lead, inorganic mercury, nickel, selenium ethyl benzene, phenol and toluene. Draft SGVs for benzene, naphthalene and xylene have been produced but their publication is on hold. Toxicological data (Tox) has been published for each of these contaminants as well as for benzo[a]pyrene, benzene, dioxins, furans and dioxin-like PCBs, naphthalene, vinyl chloride, 1,1,2,2 tetrachloroethane and 1,1,1,2 tetrachloroethane, 1,1,1 trichloroethane, tetrachloroethene, carbon tetrachloride, 1,2-dichloroethane, trichloroethene and xylene. The SGVs for ethyl benzene, phenol and toluene are dependent on the soil organic matter (SOM) content (which can be calculated from the total organic carbon (TOC) content). As an initial screen the SGVs for 1% SOM are considered to be appropriate.

India

In March 2009, the issue of Uranium poisoning in Punjab attracted press coverage. It was alleged to be caused by fly ash ponds of thermal power stations, which reportedly lead to severe birth defects in children in the Faridkot and Bhatinda districts of Punjab. The news reports claimed the uranium levels were more than 60 times the maximum safe limit. In 2012, the Government of India confirmed that the ground water in Malwa belt of Punjab has uranium metal that is 50% above the trace limits set by the United Nations' World Health Organization. Scientific studies, based on over 1000 samples from various sampling points, could not trace the source to fly ash and any sources from thermal power plants or industry as originally alleged. The study also revealed that the uranium concentration in ground water of Malwa district is not 60 times the WHO limits, but only 50% above the WHO limit in 3 locations. This highest concentration found in samples was less than those found naturally in ground waters currently used for human purposes elsewhere, such as Finland. Research is underway to identify natural or other sources for the uranium.

Mining

Surface coal mining

Sulfur miner with 90 kg of sulfur carried from the floor of the Ijen Volcano (2015)

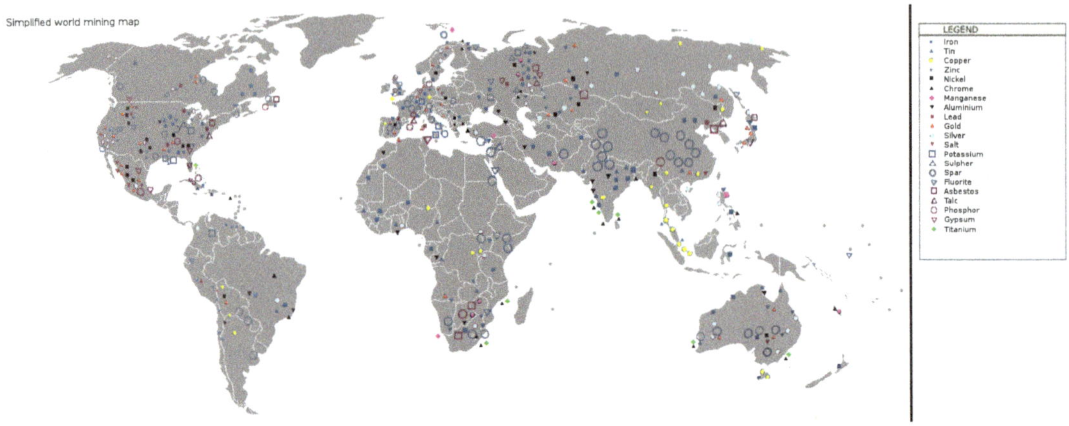

Simplified world active mining map

Mining is extraction of valuable minerals or other geological materials from the earth usually from an orebody, lode, vein, seam, reef or placer deposits. These deposits form a mineralized package that is of economic interest to the miner.

Ores recovered by mining include metals, coal, oil shale, gemstones, limestone, chalk, dimension stone, rock salt, potash, gravel, and clay. Mining is required to obtain any material that cannot be grown through agricultural processes, or created artificially in a laboratory or factory. Mining in a wider sense includes extraction of any non-renewable resource such as petroleum, natural gas, or even water.

Mining of stones and metal has been a human activity since pre-historic times. Modern mining processes involve prospecting for ore bodies, analysis of the profit potential of a proposed mine, extraction of the desired materials, and final reclamation of the land after the mine is closed.

Mining operations usually create a negative environmental impact, both during the mining activity and after the mine has closed. Hence, most of the world's nations have passed regulations to decrease the impact. Work safety has long been a concern as well, and modern practices have significantly improved safety in mines.

Levels of metals recycling are generally low. Unless future end-of-life recycling rates are stepped up, some rare metals may become unavailable for use in a variety of consumer products. Due to the low recycling rates, some landfills now contain higher concentrations of metal than mines themselves.

History

Prehistoric Mining

Since the beginning of civilization, people have used stone, ceramics and, later, metals found close to the Earth's surface. These were used to make early tools and weapons; for example, high quality flint found in northern France, southern England and Poland was used to create flint tools. Flint mines have been found in chalk areas where seams of the stone were followed underground by shafts and galleries. The mines at Grimes Graves and Krzemionki are especially famous, and like most other flint mines, are Neolithic in origin (ca 4000 BC-ca 3000 BC). Other hard rocks mined or collected for axes included the greenstone of the Langdale axe industry based in the English Lake District.

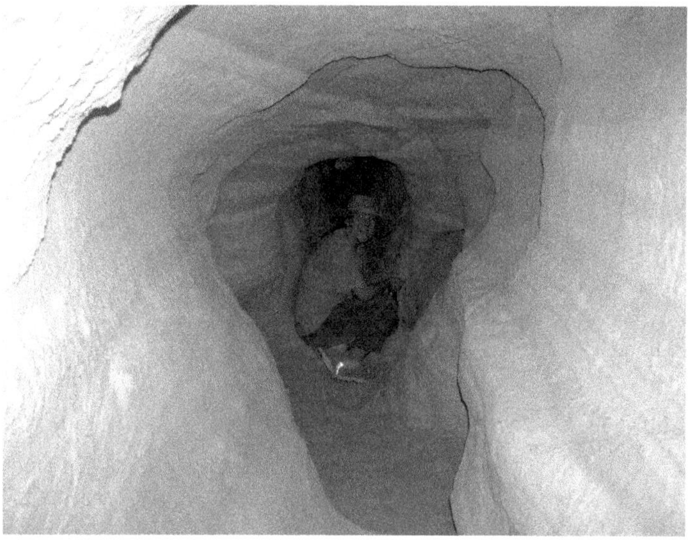

Chalcolithic copper mine in Timna Valley, Negev Desert

The oldest-known mine on archaeological record is the "Lion Cave" in Swaziland, which radiocarbon dating shows to be about 43,000 years old. At this site Paleolithic humans mined hematite to make the red pigment ochre. Mines of a similar age in Hungary are believed to be sites where Neanderthals may have mined flint for weapons and tools.

Ancient Egypt

Ancient Egyptians mined malachite at Maadi. At first, Egyptians used the bright green malachite stones for ornamentations and pottery. Later, between 2613 and 2494 BC, large building projects required expeditions abroad to the area of Wadi Maghareh in order to secure minerals and other resources not available in Egypt itself. Quarries for turquoise and copper were also found at Wadi Hammamat, Tura, Aswan and various other Nubian sites on the Sinai Peninsula and at Timna.

Mining in Egypt occurred in the earliest dynasties. The gold mines of Nubia were among the largest and most extensive of any in Ancient Egypt. These mines are described by the Greek author Diodorus Siculus, who mentions fire-setting as one method used to break down the hard rock holding the gold. One of the complexes is shown in one of the earliest known maps. The miners crushed the ore and ground it to a fine powder before washing the powder for the gold dust.

Ancient Greek and Roman Mining

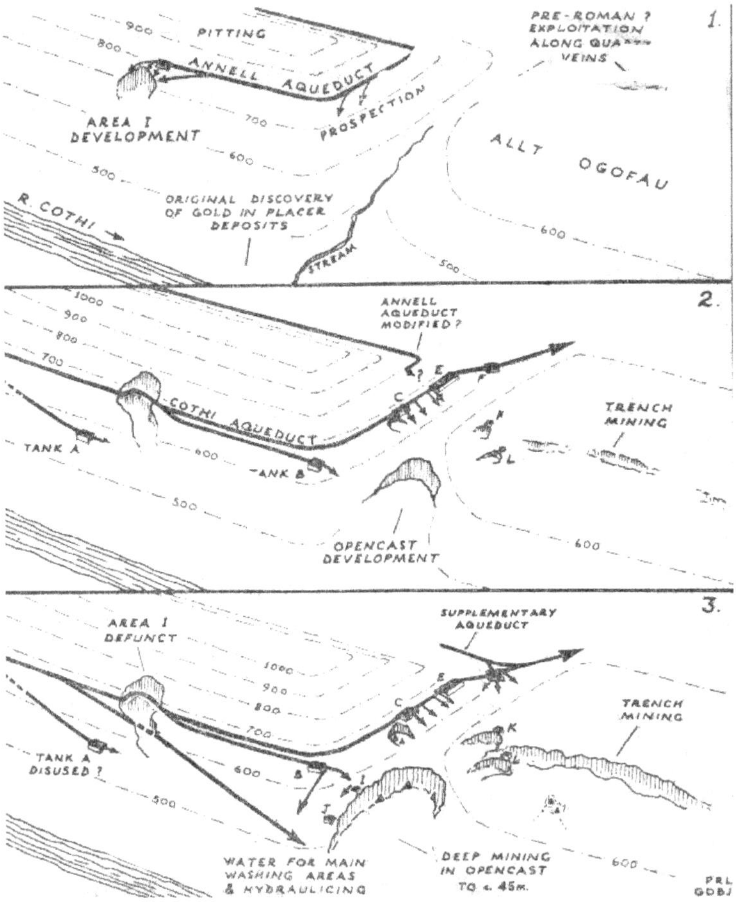

Ancient Roman development of the Dolaucothi Gold Mines, Wales

Mining in Europe has a very long history. Examples include the silver mines of Laurium, which helped support the Greek city state of Athens. Although they had over 20,000 slaves working them, their technology was essentially identical to their Bronze Age predecessors. At other mines, such as on the island of Thassos, marble was quarried by the Parians after they arrived in the 7th Century BC. The marble was shipped away and was later found by archaeologists to have been used in buildings including the tomb of Amphipolis. Philip II of Macedon, the father of Alexander the Great, captured the gold mines of Mount Pangeo in 357 BC to fund his military campaigns. He also captured gold mines in Thrace for minting coinage, eventually producing 26 tons per year.

However, it was the Romans who developed large scale mining methods, especially the use of large volumes of water brought to the minehead by numerous aqueducts. The water was used for a variety of purposes, including removing overburden and rock debris, called hydraulic mining, as well as washing comminuted, or crushed, ores and driving simple machinery.

The Romans used hydraulic mining methods on a large scale to prospect for the veins of ore, especially a now-obsolete form of mining known as hushing. They built numerous aqueducts to supply water to the minehead. There, the water stored in large reservoirs and tanks. When a full tank was opened, the flood of water sluiced away the overburden to expose the bedrock underneath and any gold veins. The rock was then worked upon by fire-setting to heat the rock, which would be quenched with a stream of water. The resulting thermal shock cracked the rock, enabling it to be removed by further streams of water from the overhead tanks. The Roman miners used similar methods to work cassiterite deposits in Cornwall and lead ore in the Pennines.

The methods had been developed by the Romans in Spain in 25 AD to exploit large alluvial gold deposits, the largest site being at Las Medulas, where seven long aqueducts tapped local rivers and sluiced the deposits. Spain was one of the most important mining regions, but all regions of the Roman Empire were exploited. In Great Britain the natives had mined minerals for millennia, but after the Roman conquest, the scale of the operations increased dramatically, as the Romans needed Britannia's resources, especially gold, silver, tin, and lead.

Roman techniques were not limited to surface mining. They followed the ore veins underground once opencast mining was no longer feasible. At Dolaucothi they stoped out the veins and drove adits through bare rock to drain the stopes. The same adits were also used to ventilate the workings, especially important when fire-setting was used. At other parts of the site, they penetrated the water table and dewatered the mines using several kinds of machines, especially reverse overshot water-wheels. These were used extensively in the copper mines at Rio Tinto in Spain, where one sequence comprised 16 such wheels arranged in pairs, and lifting water about 24 metres (79 ft). They were worked as treadmills with miners standing on the top slats. Many examples of such devices have been found in old Roman mines and some examples are now preserved in the British Museum and the National Museum of Wales.

Medieval Europe

Mining as an industry underwent dramatic changes in medieval Europe. The mining industry in the early Middle Ages was mainly focused on the extraction of copper and iron. Other precious metals were also used, mainly for gilding or coinage. Initially, many metals were obtained through open-pit mining, and ore was primarily extracted from shallow depths, rather than through deep mine shafts. Around the 14th century, the growing use of weapons, armour, stirrups, and horseshoes greatly increased the demand for iron. Medieval knights, for example, were often laden with up to 100 pounds of plate or chain link armour in addition to swords, lances and other weapons. The overwhelming dependency on iron for military purposes spurred iron production and extraction processes.

Agricola, author of *De Re Metallica*

Gallery, 12th to 13th century, Germany

The silver crisis of 1465 occurred when all mines had reached depths at which the shafts could no longer be pumped dry with the available technology. Although an increased use of bank notes, credit and copper coins during this period did decrease the value of, and dependence on, precious metals, gold and silver still remained vital to the story of medieval mining.

Due to differences in the social structure of society, the increasing extraction of mineral deposits spread from central Europe to England in the mid-sixteenth century.

On the continent, mineral deposits belonged to the crown, and this regalian right was stoutly maintained. But in England, royal mining rights were restricted to gold and silver (of which England had virtually no deposits) by a judicial decision of 1568 and a law in 1688. England had iron, zinc, copper, lead, and tin ores. Landlords who owned the base metals and coal under their estates then had a strong inducement to extract these metals or to lease the deposits and collect royalties from mine operators. English, German, and Dutch capital combined to finance extraction and refining. Hundreds of German technicians and skilled workers were brought over; in 1642 a colony of 4,000 foreigners was mining and smelting copper at Keswick in the northwestern mountains.

Use of water power in the form of water mills was extensive. The water mills were employed in crushing ore, raising ore from shafts, and ventilating galleries by powering giant bellows. Black powder was first used in mining in Selmecbánya, Kingdom of Hungary (now Banská Štiavnica, Slovakia) in 1627. Black powder allowed blasting of rock and earth to loosen and reveal ore veins. Blasting was much faster than fire-setting and allowed the mining of previously impenetrable metals and ores. In 1762, the world's first mining academy was established in the same town there.

The widespread adoption of agricultural innovations such as the iron plowshare, as well as the growing use of metal as a building material, was also a driving force in the tremendous growth of the iron industry during this period. Inventions like the arrastra were often used by the Spanish to pulverize ore after being mined. This device was powered by animals and used the same principles used for grain threshing.

Much of the knowledge of medieval mining techniques comes from books such as Biringuccio's *De la pirotechnia* and probably most importantly from Georg Agricola's *De re metallica* (1556). These books detail many different mining methods used in German and Saxon mines. A prime issue in medieval mines, which Agricola explains in detail, was the removal of water from mining shafts. As miners dug deeper to access new veins, flooding became a very real obstacle. The mining industry became dramatically more efficient and prosperous with the invention of mechanical and animal driven pumps.

Classical Philippine Civilization

Mining in the Philippines began around 1000 BC. The early Filipinos worked various mines of gold, silver, copper and iron. Jewels, gold ingots, chains, calombigas and earrings were handed down from antiquity and inherited from their ancestors. Gold dagger handles, gold dishes, tooth plating, and huge gold ornamets were also used. In Laszlo Legeza's "Tantric elements in pre-Hispanic Philippines Gold Art", he mentioned that gold jewelry of Philippine origin was found in Ancient Egypt. According to Antonio Pigafetta, the people of Mindoro possessed great skill in mixing gold with other metals and gave it a natural and perfect appearance that could deceive even the best of silversmiths. The natives were also known for the jewelries made of other precious stones such as carnelian, agate and pearl. Some outstanding examples of Philippine jewelry included necklaces, belts, armlets and rings placed around the waist.

The image of a Maharlika class of the Philippine Society , depicted in Boxer Codex that the Gold used as a form of Jewelry.

The Americas

Lead mining in the upper Mississippi River region of the U.S., 1865.

There are ancient, prehistoric copper mines along Lake Superior, and metallic copper was still found there, near the surface, in colonial times.

Indigenous peoples availed themselves of this copper starting at least 5,000 years ago," and copper tools, arrowheads, and other artifacts that were part of an extensive native trade network have been discovered. In addition, obsidian, flint, and other minerals were mined, worked, and traded. Early French explorers who encountered the sites made no use of the metals due to the difficulties of transporting them, but the copper was eventually traded throughout the continent along major river routes.

Miners at the Tamarack Mine in Copper Country, Michigan, U.S. in 1905.

In the early colonial history of the Americas, "native gold and silver was quickly expropriated and sent back to Spain in fleets of gold- and silver-laden galleons," the gold and silver originating mostly from mines in Central and South America. Turquoise dated at 700 A.D. was mined in pre-Columbian America; in the Cerillos Mining District in New Mexico, estimates are that "about 15,000 tons of rock had been removed from Mt. Chalchihuitl using stone tools before 1700."

Mining in the United States became prevalent in the 19th century, and the General Mining Act of 1872 was passed to encourage mining of federal lands. As with the California Gold Rush in the mid-19th century, mining for minerals and precious metals, along with ranching, was a driving factor in the Westward Expansion to the Pacific coast. With the exploration of the West, mining camps were established and "expressed a distinctive spirit, an enduring legacy to the new nation;" Gold Rushers would experience the same problems as the Land Rushers of the transient West that preceded them. Aided by railroads, many traveled West for work opportunities in mining. Western cities such as Denver and Sacramento originated as mining towns.

When new areas were explored, it was usually the gold (placer and then lode) and then silver that were taken into possession and extracted first. Other metals would often wait for railroads or canals, as coarse gold dust and nuggets do not require smelting and are easy to identify and transport.

Modern Period

In the early 20th century, the gold and silver rush to the western United States also stimulated mining for coal as well as base metals such as copper, lead, and iron. Areas in modern Montana, Utah, Arizona, and later Alaska became predominate suppliers of copper to the world, which was increasingly demanding copper for electrical and households goods. Canada's mining industry grew more slowly than did the United

States' due to limitations in transportation, capital, and U.S. competition; Ontario was the major producer of the early 20th century with nickel, copper, and gold.

View showing miners' clothes suspended by pulleys, also wash basins and ventilation system, Kirkland Lake, Ontario, 1936.

Meanwhile, Australia experienced the Australian gold rushes and by the 1850s was producing 40% of the world's gold, followed by the establishment of large mines such as the Mount Morgan Mine, which ran for nearly a hundred years, Broken Hill ore deposit (one of the largest zinc-lead ore deposits), and the iron ore mines at Iron Knob. After declines in production, another boom in mining occurred in the 1960s. Now, in the early 21st century, Australia remains a major world mineral producer.

As the 21st century begins, a globalized mining industry of large multinational corporations has arisen. Peak minerals and environmental impacts have also become a concern. Different elements, particularly rare earth minerals, have begun to increase in demand as a result of new technologies.

Mine Development and Lifecycle

The process of mining from discovery of an ore body through extraction of minerals and finally to returning the land to its natural state consists of several distinct steps. The first is discovery of the ore body, which is carried out through prospecting or exploration to find and then define the extent, location and value of the ore body. This leads to a mathematical resource estimation to estimate the size and grade of the deposit.

This estimation is used to conduct a pre-feasibility study to determine the theoretical economics of the ore deposit. This identifies, early on, whether further investment in estimation and engineering studies is warranted and identifies key risks and areas for further work. The next step is to conduct a feasibility study to evaluate the financial viability, the technical and financial risks, and the robustness of the project.

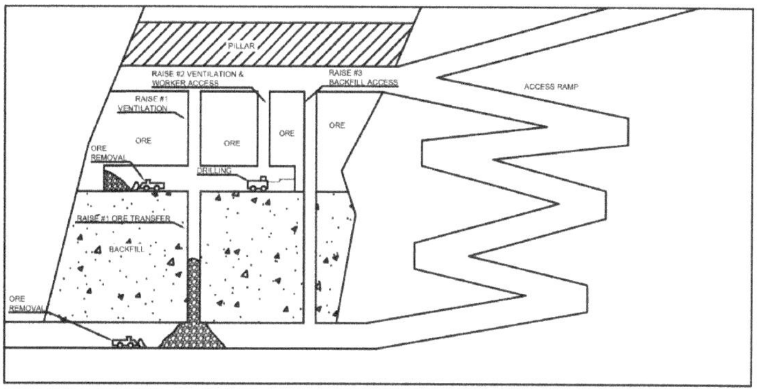

Schematic of a cut and fill mining operation in hard rock.

This is when the mining company makes the decision whether to develop the mine or to walk away from the project. This includes mine planning to evaluate the economically recoverable portion of the deposit, the metallurgy and ore recoverability, marketability and payability of the ore concentrates, engineering concerns, milling and infrastructure costs, finance and equity requirements, and an analysis of the proposed mine from the initial excavation all the way through to reclamation. The proportion of a deposit that is economically recoverable is dependent on the enrichment factor of the ore in the area.

To gain access to the mineral deposit within an area it is often necessary to mine through or remove waste material which is not of immediate interest to the miner. The total movement of ore and waste constitutes the mining process. Often more waste than ore is mined during the life of a mine, depending on the nature and location of the ore body. Waste removal and placement is a major cost to the mining operator, so a detailed characterization of the waste material forms an essential part of the geological exploration program for a mining operation.

Once the analysis determines a given ore body is worth recovering, development begins to create access to the ore body. The mine buildings and processing plants are built, and any necessary equipment is obtained. The operation of the mine to recover the ore begins and continues as long as the company operating the mine finds it economical to do so. Once all the ore that the mine can produce profitably is recovered, reclamation begins to make the land used by the mine suitable for future use.

Mining Techniques

Mining techniques can be divided into two common excavation types: surface mining and sub-surface (underground) mining. Today, surface mining is much more common, and produces, for example, 85% of minerals (excluding petroleum and natural gas) in the United States, including 98% of metallic ores.

Underground longwall mining.

Targets are divided into two general categories of materials: *placer deposits*, consisting of valuable minerals contained within river gravels, beach sands, and other unconsolidated materials; and *lode deposits*, where valuable minerals are found in veins, in layers, or in mineral grains generally distributed throughout a mass of actual rock. Both types of ore deposit, placer or lode, are mined by both surface and underground methods.

Some mining, including much of the rare earth elements and uranium mining, is done by less-common methods, such as in-situ leaching: this technique involves digging neither at the surface nor underground. The extraction of target minerals by this technique requires that they be soluble, e.g., potash, potassium chloride, sodium chloride, sodium sulfate, which dissolve in water. Some minerals, such as copper minerals and uranium oxide, require acid or carbonate solutions to dissolve.

Surface Mining

Surface mining is done by removing (stripping) surface vegetation, dirt, and, if necessary, layers of bedrock in order to reach buried ore deposits. Techniques of surface mining include: open-pit mining, which is the recovery of materials from an open pit in the ground, quarrying, identical to open-pit mining except that it refers to sand, stone and clay; strip mining, which consists of stripping surface layers off to reveal ore/ seams underneath; and mountaintop removal, commonly associated with coal mining, which involves taking the top of a mountain off to reach ore deposits at depth. Most (but not all) placer deposits, because of their shallowly buried nature, are mined by surface methods. Finally, landfill mining involves sites where landfills are excavated and processed.

Garzweiler surface mine, Germany

Underground Mining

Mantrip used for transporting miners within an underground mine

Sub-surface mining consists of digging tunnels or shafts into the earth to reach buried ore deposits. Ore, for processing, and waste rock, for disposal, are brought to the surface through the tunnels and shafts. Sub-surface mining can be classified by the type of access shafts used, the extraction method or the technique used to reach the mineral deposit. Drift mining utilizes horizontal access tunnels, slope mining uses diagonally sloping access shafts, and shaft mining utilizes vertical access shafts. Mining in hard and soft rock formations require different techniques.

Other methods include shrinkage stope mining, which is mining upward, creating a sloping underground room, long wall mining, which is grinding a long ore surface underground, and room and pillar mining, which is removing ore from rooms while leaving pillars in place to support the roof of the room. Room and pillar mining often leads to retreat mining, in which supporting pillars are removed as miners retreat, allowing the room to cave in, thereby loosening more ore. Additional sub-surface mining methods include hard rock mining, which is mining of hard rock (igneous, metamorphic or sedimentary) materials, bore hole mining, drift and fill mining, long hole slope mining, sub level caving, and block caving.

Highwall Mining

Caterpillar Highwall Miner HW300 - Technology Bridging Underground and Open Pit Mining

Highwall mining is another form of surface mining that evolved from auger mining. In Highwall mining, the coal seam is penetrated by a continuous miner propelled by a hydraulic Pushbeam Transfer Mechanism (PTM). A typical cycle includes sumping (launch-pushing forward) and shearing (raising and lowering the cutterhead boom to cut the entire height of the coal seam). As the coal recovery cycle continues, the cutterhead is progressively launched into the coal seam for 19.72 feet (6.01 m). Then, the Pushbeam Transfer Mechanism (PTM) automatically inserts a 19.72-foot (6.01 m) long rectangular Pushbeam (Screw-Conveyor Segment) into the center section of the machine between the Powerhead and the cutterhead. The Pushbeam system can penetrate nearly 1,000 feet (300 m) into the coal seam. One patented Highwall mining system uses augers enclosed inside the Pushbeam that prevent the mined coal from being contaminated by rock debris during the conveyance process. Using a video imaging and/or a gamma ray sensor and/or other Geo-Radar systems like a coal-rock interface detection sensor (CID), the operator can see ahead projection of the seam-rock interface and guide the continuous miner's progress. Highwall mining can produce thousands of tons of coal in contour-strip operations with narrow benches, previously mined areas, trench mine applications and steep-dip seams with controlled water-inflow pump system and/or a gas (inert) venting system.

Machines

Heavy machinery is used in mining to explore and develop sites, to remove and stockpile overburden, to break and remove rocks of various hardness and toughness, to process the ore, and to carry out reclamation projects after the mine is closed. Bulldozers, drills, explosives and trucks are all necessary for excavating the land. In the case of placer mining, unconsolidated gravel, or alluvium, is fed into machinery consisting of a hopper and a shaking screen or trommel which frees the desired minerals from the waste gravel. The minerals are then concentrated using sluices or jigs.

The Bagger 288 is a bucket-wheel excavator used in strip mining. It is also the largest land vehicle of all time.

A Bucyrus Erie 2570 dragline and CAT 797 haul truck at the North Antelope Rochelle opencut coal mine

Large drills are used to sink shafts, excavate stopes, and obtain samples for analysis. Trams are used to transport miners, minerals and waste. Lifts carry miners into and out of mines, and move rock and ore out, and machinery in and out, of underground mines. Huge trucks, shovels and cranes are employed in surface mining to move large quantities of overburden and ore. Processing plants utilize large crushers, mills, reactors, roasters and other equipment to consolidate the mineral-rich material and extract the desired compounds and metals from the ore.

Processing

Once the mineral is extracted, it is often then processed. The science of extractive metallurgy is a specialized area in the science of metallurgy that studies the extraction of valuable metals from their ores, especially through chemical or mechanical means.

Mineral processing (or mineral dressing) is a specialized area in the science of metallurgy that studies the mechanical means of crushing, grinding, and washing that enable the separation (extractive metallurgy) of valuable metals or minerals from their gangue (waste material). Processing of placer ore material consists of gravity-dependent methods of separation, such as sluice boxes. Only minor shaking or washing may be necessary to disaggregate (unclump) the sands or gravels before processing. Processing of ore from a lode mine, whether it is a surface or subsurface mine, requires that the rock ore be crushed and pulverized before extraction of the valuable minerals begins. After lode ore is crushed, recovery of the valuable minerals is done by one, or a combination of several, mechanical and chemical techniques.

Since most metals are present in ores as oxides or sulfides, the metal needs to be reduced to its metallic form. This can be accomplished through chemical means such as smelting or through electrolytic reduction, as in the case of aluminium. Geometallurgy combines the geologic sciences with extractive metallurgy and mining.

Environmental Effects

Iron hydroxide precipitate stains a stream receiving acid drainage from surface coal mining.

Environmental issues can include erosion, formation of sinkholes, loss of biodiversity, and contamination of soil, groundwater and surface water by chemicals from mining processes. In some cases, additional forest logging is done in the vicinity of mines to create space for the storage of the created debris and soil. Contamination resulting from leakage of chemicals can also affect the health of the local population if not properly controlled. Extreme examples of pollution from mining activities include coal fires, which can last for years or even decades, producing massive amounts of environmental damage.

Mining companies in most countries are required to follow stringent environmental and rehabilitation codes in order to minimize environmental impact and avoid impacting human health. These codes and regulations all require the common steps of environmental impact assessment, development of environmental management plans, mine closure planning (which must be done before the start of mining operations), and environmental monitoring during operation and after closure. However, in some areas, particularly in the developing world, government regulations may not be well enforced.

For major mining companies and any company seeking international financing, there are a number of other mechanisms to enforce good environmental standards. These generally relate to financing standards such as the Equator Principles, IFC environmental standards, and criteria for Socially responsible investing. Mining companies have used this oversight from the financial sector to argue for some level of industry self-regulation. In 1992, a Draft Code of Conduct for Transnational Corporations was proposed at the Rio Earth Summit by the UN Centre for Transnational Corporations (UNCTC), but the Business Council for Sustainable Development (BCSD) together with the International Chamber of Commerce (ICC) argued successfully for self-regulation instead.

This was followed by the Global Mining Initiative which was begun by nine of the largest metals and mining companies and which led to the formation of the International Council on Mining and Metals, whose purpose was to "act as a catalyst" in an effort to improve social and environmental performance in the mining and metals industry internationally. The mining industry has provided funding to various conservation groups, some of which have been working with conservation agendas that are at odds with an emerging acceptance of the rights of indigenous people – particularly the right to make land-use decisions.

Certification of mines with good practices occurs through the International Organization for Standardization (ISO). For example, ISO 9000 and ISO 14001, which certify an "auditable environmental management system", involve short inspections, although they have been accused of lacking rigor. Certification is also available through Ceres' Global Reporting Initiative, but these reports are voluntary and unverified. Miscellaneous other certification programs exist for various projects, typically through nonprofit groups.

The purpose of a 2012 EPS PEAKS paper was to provide evidence on policies managing ecological costs and maximise socio-economic benefits of mining using host country regulatory initiatives. It found existing literature suggesting donors encourage developing countries to:

- Make the environment-poverty link and introduce cutting-edge wealth measures and natural capital accounts.
- Reform old taxes in line with more recent financial innovation, engage directly

with the companies, enacting land use and impact assessments, and incorporate specialised support and standards agencies.

- Set in play transparency and community participation initiatives using the wealth accrued.

Waste

Ore mills generate large amounts of waste, called tailings. For example, 99 tons of waste are generated per ton of copper, with even higher ratios in gold mining - because only 5.3 g of gold is extracted per ton of ore, a ton of gold produces 200,000 tons of tailings. (As time goes on and richer deposits are exhausted - and technology improves to permit - this number is going down to .5 g and less.) These tailings can be toxic. Tailings, which are usually produced as a slurry, are most commonly dumped into ponds made from naturally existing valleys. These ponds are secured by impoundments (dams or embankment dams). In 2000 it was estimated that 3,500 tailings impoundments existed, and that every year, 2 to 5 major failures and 35 minor failures occurred; for example, in the Marcopper mining disaster at least 2 million tons of tailings were released into a local river. Subaqueous tailings disposal is another option. The mining industry has argued that submarine tailings disposal (STD), which disposes of tailings in the sea, is ideal because it avoids the risks of tailings ponds; although the practice is illegal in the United States and Canada, it is used in the developing world.

The waste is classified as either sterile or mineralised, with acid generating potential, and the movement and storage of this material forms a major part of the mine planning process. When the mineralised package is determined by an economic cut-off, the near-grade mineralised waste is usually dumped separately with view to later treatment should market conditions change and it becomes economically viable. Civil engineering design parameters are used in the design of the waste dumps, and special conditions apply to high-rainfall areas and to seismically active areas. Waste dump designs must meet all regulatory requirements of the country in whose jurisdiction the mine is located. It is also common practice to rehabilitate dumps to an internationally acceptable standard, which in some cases means that higher standards than the local regulatory standard are applied.

Renewable Energy and Mining

Many mining sites are remote and not connected to the grid. Electricity is typically generated with diesel generators. Due to high transportation cost and theft during transportation the cost for generating electricity is normally high. Renewable energy applications are becoming an alternative or amendment. Both solar and wind power plants can contribute in saving diesel costs at mining sites. Renewable energy applications have been built at mining sites. Cost savings can reach up to 70%.

Mining Industry

Mining exists in many countries. London is known as the capital of global "mining houses" such as Rio Tinto Group, BHP Billiton, and Anglo American PLC. The US mining industry is also large, but it is dominated by the coal and other nonmetal minerals (e.g., rock and sand), and various regulations have worked to reduce the significance of mining in the United States. In 2007 the total market capitalization of mining companies was reported at US$962 billion, which compares to a total global market cap of publicly traded companies of about US$50 trillion in 2007. In 2002, Chile and Peru were reportedly the major mining countries of South America. The mineral industry of Africa includes the mining of various minerals; it produces relatively little of the industrial metals copper, lead, and zinc, but according to one estimate has as a percent of world reserves 40% of gold, 60% of cobalt, and 90% of the world's platinum group metals. Mining in India is a significant part of that country's economy. In the developed world, mining in Australia, with BHP Billiton founded and headquartered in the country, and mining in Canada are particularly significant. For rare earth minerals mining, China reportedly controlled 95% of production in 2013.

The Bingham Canyon Mine of Rio Tinto's subsidiary, Kennecott Utah Copper.

While exploration and mining can be conducted by individual entrepreneurs or small businesses, most modern-day mines are large enterprises requiring large amounts of capital to establish. Consequently, the mining sector of the industry is dominated by large, often multinational, companies, most of them publicly listed. It can be argued that what is referred to as the 'mining industry' is actually two sectors, one specializing in exploration for new resources and the other in mining those resources. The exploration sector is typically made up of individuals and small mineral resource companies, called "juniors", which are dependent on venture capital. The mining sector is made up of large multinational companies that are sustained by production from their mining operations. Various other industries such as equipment manufacture, environmental testing, and metallurgy analysis rely on, and support, the mining industry throughout the world. Canadian stock exchanges have a particular focus on mining companies, particularly junior exploration companies through Toronto's TSX Venture Exchange;

Canadian companies raise capital on these exchanges and then invest the money in exploration globally. Some have argued that below juniors there exists a substantial sector of illegitimate companies primarily focused on manipulating stock prices.

Mining operations can be grouped into five major categories in terms of their respective resources. These are oil and gas extraction, coal mining, metal ore mining, nonmetallic mineral mining and quarrying, and mining support activities. Of all of these categories, oil and gas extraction remains one of the largest in terms of its global economic importance. Prospecting potential mining sites, a vital area of concern for the mining industry, is now done using sophisticated new technologies such as seismic prospecting and remote-sensing satellites. Mining is heavily affected by the prices of the commodity minerals, which are often volatile. The 2000s commodities boom ("commodities supercycle") increased the prices of commodities, driving aggressive mining. In addition, the price of gold increased dramatically in the 2000s, which increased gold mining; for example, one study found that conversion of forest in the Amazon increased six-fold from the period 2003–2006 (292 ha/yr) to the period 2006–2009 (1,915 ha/yr), largely due to artisanal mining.

Corporate Classifications

Mining companies can be classified based on their size and financial capabilities:

- Major companies are considered to have an adjusted annual mining-related revenue of more than US$500 million, with the financial capability to develop a major mine on its own.
- Intermediate companies have at least $50 million in annual revenue but less than $500 million.
- Junior companies rely on equity financing as their principal means of funding exploration. Juniors are mainly pure exploration companies, but may also produce minimally, and do not have a revenue exceeding US$50 million.

Regulation and Governance

New regulations and a process of legislative reforms aim to improve the harmonization and stability of the mining sector in mineral-rich countries. New legislation for mining industry in African countries still appears to be an issue, but has the potential to be solved, when a consensus is reached on the best approach. By the beginning of the 21st century the booming and increasingly complex mining sector in mineral-rich countries was providing only slight benefits to local communities, especially in given the sustainability issues. Increasing debate and influence by NGOs and local communities called for a new approahes which would also include disadvantaged communities, and work towards sustainable development even after mine closure (including transparency and revenue management). By the early 2000s, community development issues and resettlements became mainstream concerns in World Bank mining projects. Mining-industry expansion after mineral prices increased in 2003 and also potential fiscal revenues in those countries created an omission in the other economic sectors in terms of finances and development. Furthermore, this highlighted regional and local demand for mining revenues and an inability of sub-national governments to effectively use the revenues. The Fraser Institute (a Canadian think tank) has highlighted the environmental protection laws in developing countries, as well as voluntary efforts by mining companies to improve their environmental impact.

In 2007 the Extractive Industries Transparency Initiative (EITI) was mainstreamed in all countries cooperating with the World Bank in mining industry reform. The EITI operates and was implemented with the support of the EITI multi-donor trust fund, managed by the World Bank. The Extractive Industries Transparency Initiative (EITI) aims to increase transparency in transactions between governments and companies in extractive industries by monitoring the revenues and benefits between industries and recipient governments. The entrance process is voluntary for each country and is monitored by multiple stakeholders including governments, private companies and civil society representatives, responsible for disclosure and dissemination of the reconciliation report; however, the competitive disadvantage of company-by company public report is for some of the businesses in Ghana at least, the main constraint. Therefore, the outcome assessment in terms of failure or success of the new EITI regulation does not only "rest on the government's shoulders" but also on civil society and companies.

On the other hand, implementation has issues; inclusion or exclusion of artisanal mining and small-scale mining (ASM) from the EITI and how to deal with "non-cash" payments made by companies to subnational governments. Furthermore, the disproportionate revenues the mining industry can bring to the comparatively small number of people that it employs, causes other problems, like a lack of investment in other less lucrative sectors, leading to swings in government revenuebecause of volatility in the oil markets. Artisanal mining is clearly an issue in EITI Countries such as the Central African Republic, D.R. Congo, Guinea, Liberia and Sierra Leone – i.e. almost half of the mining countries implementing the EITI. Among other things, limited scope of the EITI involving disparity in terms of knowledge of the industry and negotiation skills, thus far flexibility of the policy (e.g. liberty of the countries to expand beyond the minimum requirements and adapt it to their needs), creates another risk of unsuccessful implementation. Public awareness increase, where government should act as a bridge between public and initiative for a successful outcome of the policy is an important element to be considered.

World Bank

The World Bank has been involved in mining since 1955, mainly through grants from its International Bank for Reconstruction and Development, with the Bank's Multilateral Investment Guarantee Agency offering political risk insurance. Between 1955 and 1990 it provided about $2 billion to fifty mining projects, broadly categorized as reform and rehabilitation, greenfield mine construction, mineral processing, technical assistance, and engineering. These projects have been criticized, particularly the Ferro Carajas project of Brazil, begun in 1981. The World Bank established mining codes intended to increase foreign investment; in 1988 it solicited feedback from 45 mining companies on how to increase their involvement.

In 1992 the World Bank began to push for privatization of government-owned mining companies with a new set of codes, beginning with its report *The Strategy for African Mining*. In 1997, Latin America's largest miner Companhia Vale do Rio Doce (CVRD) was privatized. These and other developments such as the Philippines 1995 Mining Act led the bank to publish a third report (*Assistance for Minerals Sector Development and Reform in Member Countries*) which endorsed mandatory environment impact assessments and attention to the concerns of the local population. The codes based on this report are influential in the legislation of developing nations. The new codes are intended to encourage development through tax holidays, zero custom duties, reduced income taxes, and related measures. The results of these codes were analyzed by a group from the University of Quebec, which concluded that the codes promote foreign investment but "fall very short of permitting sustainable development". The observed negative correlation between natural resources and economic development is known as the resource curse.

Safety

Mining transport in Devnya, Bulgaria.

Safety has long been a concern in the mining business, especially in sub-surface mining. The Courrières mine disaster, Europe's worst mining accident, involved the death of 1,099 miners in Northern France on March 10, 1906. This disaster was surpassed only by the Benxihu Colliery accident in China on April 26, 1942, which killed 1,549 miners. While mining today is substantially safer than it was in previous decades, mining accidents still occur. Government figures indicate that 5,000 Chinese miners die in accidents each year, while other reports have suggested a figure as high as 20,000. Mining accidents continue worldwide, including accidents causing dozens of fatalities at a time such as the 2007 Ulyanovskaya Mine disaster in Russia, the 2009 Heilongjiang mine explosion in China, and the 2010 Upper Big Branch Mine disaster in the United States.

There are numerous occupational hazards associated with mining, including exposure to rockdust which can lead to diseases such as silicosis, asbestosis, and pneumoconiosis. Gases in the mine can lead to asphyxiation and could also be ignited. Mining equipment can generate considerable noise, putting workers at risk for hearing loss. Cave-ins, rock falls, and exposure to excess heat are also known hazards.

Proper ventilation, hearing protection, and spraying equipment with water are important safety practices in mines.

Records

Chuquicamata, Chile, site of the largest circumference and second deepest open pit copper mine in the world.

As of 2008, the deepest mine in the world is TauTona in Carletonville, South Africa at 3.9 kilometres (2.4 mi), replacing the neighboring Savuka Mine in the North West Province of South Africa at 3,774 metres (12,382 ft). East Rand Mine in Boksburg, South Africa briefly held the record at 3,585 metres (11,762 ft), and the first mine declared the deepest in the world was also TauTona when it was at 3,581 metres (11,749 ft).

The Moab Khutsong gold mine in North West Province (South Africa) has the world's longest winding steel wire rope, able to lower workers to 3,054 metres (10,020 ft) in one uninterrupted four-minute journey.

The deepest mine in Europe is the 16th shaft of the uranium mines in Příbram, Czech Republic at 1,838 metres (6,030 ft), second is Bergwerk Saar in Saarland, Germany at 1,750 metres (5,740 ft).

The deepest open-pit mine in the world is Bingham Canyon Mine in Bingham Canyon, Utah, United States at over 1,200 metres (3,900 ft). The largest and second deepest open-pit copper mine in the world is Chuquicamata in Chuquicamata, Chile at 900 metres (3,000 ft), 443,000 tons of copper and 20,000 tons of molybdenum produced annually.

The deepest open-pit mine with respect to sea level is Tagebau Hambach in Germany, where the base of the pit is 293 metres (961 ft) below sea level.

The largest underground mine is Kiirunavaara Mine in Kiruna, Sweden. With 450 kilometres (280 mi) of roads, 40 million tonnes of ore produced yearly, and a depth of 1,270 metres (4,170 ft), it is also one of the most modern underground mines. The deepest borehole in the world is Kola Superdeep Borehole at 12,262 metres (40,230 ft). This, however, is not a matter of mining but rather related to scientific drilling.

Metal Reserves and Recycling

During the 20th century, the variety of metals used in society grew rapidly. Today, the development of major nations such as China and India and advances in technologies are fueling an ever-greater demand. The result is that metal mining activities are expanding and more and more of the world's metal stocks are above ground in use rather than below ground as unused reserves. An example is the in-use stock of copper. Between 1932 and 1999, copper in use in the USA rose from 73 kilograms (161 lb) to 238 kilograms (525 lb) per person.

95% of the energy used to make aluminium from bauxite ore is saved by using recycled material. However, levels of metals recycling are generally low. In 2010, the International Resource Panel, hosted by the United Nations Environment Programme (UNEP), published reports on metal stocks that exist within society and their recycling rates.

The report's authors observed that the metal stocks in society can serve as huge mines above ground. However, they warned that the recycling rates of some rare metals used in applications such as mobile phones, battery packs for hybrid cars, and fuel cells are so low that unless future end-of-life recycling rates are dramatically stepped up these critical metals will become unavailable for use in modern technology.

As recycling rates are low and so much metal has already been extracted, some landfills now contain a higher concentrations of metal than mines themselves. This is especially true of aluminium, used in cans, and precious metals, found in discarded electronics. Furthermore, waste after 15 years has still not broken down, so less processing would be required when compared to mining ores. A study undertaken by Cranfield University has found £360 million of metals could be mined from just 4 landfill sites. There is also up to 20MJ/kg of energy in waste, potentially making the re-extraction more profitable. However, although the first landfill mine opened in Tel Aviv, Israel in 1953, little work has followed due to the abundance of accessible ores.

Pore Space in Soil

The pore space of soil contains the liquid and gas phases of soil, i.e., everything but the solid phase that contains mainly minerals of varying sizes as well as organic compounds.

In order to understand porosity better a series of equations have been used to express the quantitative interactions between the three phases of soil.

Macropores or fractures play a major role in infiltration rates in many soils as well as preferential flow patterns, hydraulic conductivity and evapotranspiration. Cracks are also very influential in gas exchange, influencing respiration within soils. Modeling cracks therefore helps understand how these processes work and what the effects of changes in soil cracking such as compaction, can have on these processes.

Background

Bulk Density

$$\rho = \frac{M_s}{V_t}$$

The bulk density of soil depends greatly on the mineral make up of soil and the degree of compaction. The density of quartz is around 2.65 g/cm^3 but the bulk density of a soil may be less than half that density.

Most soils have a bulk density between 1.0 and 1.6 g/cm^3 but organic soil and some friable clay may have a bulk density well below 1 g/cm^3

Core samples are taken by driving a metal core into the earth at the desired depth and soil horizon. The samples are then oven dried and weighed.

Bulk density = (mass of oven dry soil)/volume

The bulk density of soil is inversely related to the porosity of the same soil. The more pore space in a soil the lower the value for bulk density.

Porosity

$$f = \frac{V_f}{V_t} \text{ or } f = \frac{V_a + V_w}{V_s + V_a + V_w}$$

Porosity is a measure of the total pore space in the soil. This is measured as a volume or percent. The amount of porosity in a soil depends on the minerals that make up the soil and the amount of sorting that occurs within the soil structure. For example, a sandy soil will have larger porosity than silty sand, because the silt will fill in the gaps between the sand particles.

Pore Space Relations

Hydraulic Conductivity

Hydraulic conductivity (K) is a property of soil that describes the ease with which water can move through pore spaces. It depends on the permeability of the material (pores, compaction) and on the degree of saturation. Saturated hydraulic conductivity, K_{sat}, describes water movement through saturated media. Where hydraulic conductivity has the capability to be measured at any state. It can be estimated by numerous kinds of equipment. To calculate hydraulic conductivity, Darcy's law is used. The manipulation of the law depends on the Soil saturation and instrument used.

Infiltration

Infiltration is the process by which water on the ground surface enters the soil. The water enters the soil through the pores by the forces of gravity and capillary action. The largest cracks and pores offer a great reservoir for the initial flush of water. This allows a rapid infiltration. The smaller pores take longer to fill and rely on capillary forces as well as gravity. The smaller pores have a slower infiltration as the soil becomes more saturated.

Pore Types

A pore is not simply a void in the solid structure of soil. The various pore size categories have different characteristics and contribute different attributes to soils depending on the number and frequency of each type. A widely used classification of pore size is that of Brewer (1964):

Macropore

The pores that are too large to have any significant capillary force. Unless impeded, water will drain from these pores, and they are generally air-filled at field capacity. Macropores can be caused by cracking, division of peds and aggregates, as well as plant roots, and zoological exploration. Size >75 μm.

Mesopore

The largest pores filled with water at field capacity. Also known as storage pores because of the ability to store water useful to plants. They do not have capillary forces too great so that the water does not become limiting to the plants. The properties of mesopores are highly studied by soil scientists because of their impact on agriculture and irrigation. Size 30 μm–75 μm.

Micropore

These are "pores that are sufficiently small that water within these pores is considered immobile, but available for plant extraction." Because there is little movement of water in these pores, solute movement is mainly by the process of diffusion. Size 5-30 μm.

Ultramicropore

These pores are suitable for habitation by microorganisms. Their distribution is determined by soil texture and soil organic matter, and they are not greatly affected by compaction Size 0.1-30 μm.

Cryptopore

Pores that are too small to be penetrated by most microorganisms. Organic matter in these pores is therefore protected from microbial decomposition. They are filled with water unless the soil is very dry, but little of this water is available to plants, and water movement is very slow. Size <0.1 μm.

Modelling Methods

Basic crack modeling has been undertaken for many years by simple observations and measurements of crack size, distribution, continuity and depth. These observations have either been surface observation or done on profiles in pits. Hand tracing and measurement of crack patterns on paper was one method used prior to advances in modern technology. Another field method was with the use of string and a semicircle of wire. The semi circle was moved along alternating sides of a string line. The cracks within the semicircle were measured for width, length and depth using a ruler. The crack distribution was calculated using the principle of Buffon's needle.

Disc Permeameter

This method relies on the fact that crack sizes have a range of different water potentials. At zero water potential at the soil surface an estimate of saturated hydraulic conductivity is produced, with all pores filled with water. As the potential is decreased progressively larger cracks drain. By measuring at the hydraulic conductivity at a range of negative potentials, the pore size distribution can be determined. While this is not a physical model of the cracks, it does give an indication to the sizes of pores within the soil.

Horgan and Young Model

Horgan and Young (2000) produced a computer model to create a two-dimensional prediction of surface crack formation. It used the fact that once cracks come within a certain distance of one another they tend to be attracted to each other. Cracks also tend to turn within a particular range of angles and at some stage a surface aggregate gets to a size that no more cracking will occur. These are often characteristic of a soil and can therefore be measured in the field and used in the model. However it was not able to predict the points at which cracking starts and although random in the formation of crack pattern, in many ways, cracking of soil is often not random, but follows lines of weaknesses.

Araldite-impregnation Imaging

A large core sample is collected. This is then impregnated with araldite and a fluorescent resin. The core is then cut back using a grinding implement, very gradually (~1 mm per time), and at every interval the surface of the core sample is digitally imaged. The images are then loaded into a computer where they can be analysed. Depth, continuity, surface area and a number of other measurements can then be made on the cracks within the soil.

Electrical Resistivity Imaging

Using the infinite resistivity of air, the air spaces within a soil can be mapped. A specially designed resistivity meter had improved the meter-soil contact and therefore the area of the reading. This technology can be used to produce images that can be analysed for a range of cracking properties.

Soil Permeability

Pressure, Elevation and Total Heads

In soils, the interconnected pores provide passage for water. A large number of such flow paths act together, and the average rate of flow is termed the coefficient of permeability, or just permeability. It is a measure of the ease that the soil provides to the flow of water through its pores.

The height of the water column is the pressure head (h_w).

$$\mathbf{h_w = u/\gamma_w}$$

To identify any difference in pore water pressure at different points, it is necessary to eliminate the effect of the points of measurement. With this in view, a datum is required from which locations are measured.

The elevation head (h_z) of any point is its height above the datum line. The height of water level in the standpipe above the datum is the piezometric head (h).

$$h = h_z + h_w$$

Total head consists of *three components:* elevation head, pressure head, and velocity head. As seepage velocity in soils is normally low, velocity head is ignored, and total head becomes equal to the piezometric head. Due to the low seepage velocity and small size of pores, the flow of water in the pores is steady and laminar in most cases. Water flow takes place between two points in soil due to the difference in total heads.

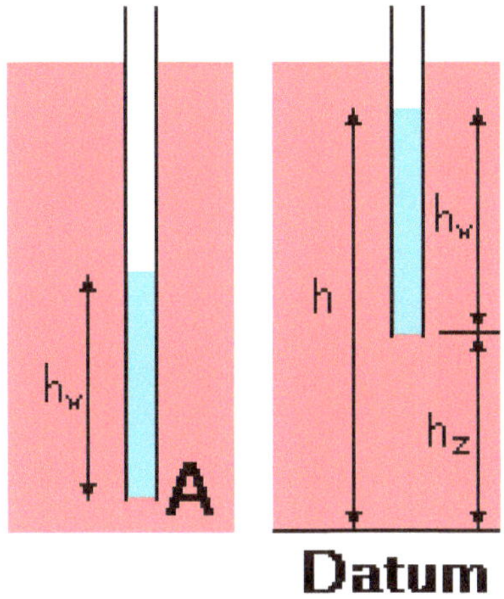

At point A, the pore water pressure (u) can be measured from the height of water in a standpipe located at that point.

Darcy's Law

Darcy's law is an equation that describes the flow of a fluid through a porous medium. The law was formulated by Henry Darcy based on the results of experiments on the flow of water through beds of sand, forming the basis of hydrogeology, a branch of earth sciences.

Background

Although Darcy's law (an expression of Newton's second law) was determined experimentally by Darcy, it has since been derived from the Navier-Stokes equations via homogenization. It is analogous to Fourier's law in the field of heat conduction, Ohm's law in the field of electrical networks, or Fick's law in diffusion theory.

One application of Darcy's law is to analyze water flow through an aquifer; Darcy's law along with the equation of conservation of mass are equivalent to the groundwater flow equation, one of the basic relationships of hydrogeology.

Morris Muskat first refined Darcy's equation for single phase flow by including viscosity in the single (fluid) phase equation of Darcy, and this change made it suitable for the petroleum industry. Based on experimental results worked out by his colleagues Wyckoff and Botset, Muskat and Meres also generalized Darcy's law to cover multiphase flow of water, oil and gas in the porous medium of a petroleum reservoir. The generalized multiphase flow equations of Muskat et alios provide the analytical foundation for reservoir engineering that exists to this day.

Description

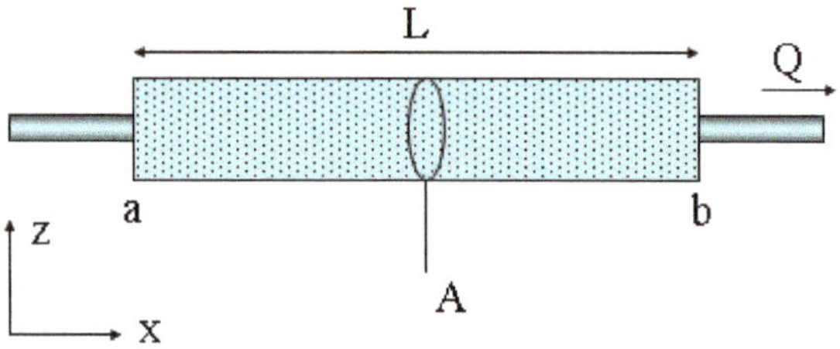

Diagram showing definitions and directions for Darcy's law.

Darcy's law, as refined by Morris Muskat, at constant elevation is a simple proportional relationship between the instantaneous discharge rate through a porous medium, the viscosity of the fluid and the pressure drop over a given distance.

$$Q = -\frac{\kappa A\left(p_{\mathrm{b}} - p_{\mathrm{a}}\right)}{\mu L}.$$

The above equation for single phase (fluid) flow is the defining equation for absolute permeability (single phase permeability). The total discharge, Q (units of volume per time, e.g., m³/s) is equal to the product of the intrinsic permeability of the medium, κ (m²), the cross-sectional area to flow, A (units of area, e.g., m²), and the total pressure drop $p_{\mathrm{b}} - p_{\mathrm{a}}$ (pascals), all divided by the viscosity, μ (Pa·s) and the length over which the pressure drop is taking place (L). The negative sign is needed because fluid flows from high pressure to low pressure. Note that the elevation head must be taken into account if the inlet and outlet are at different elevations. If the change in pressure is negative (where $p_{\mathrm{a}} > p_{\mathrm{b}}$), then the flow will be in the positive x direction. There have been several proposals for a constitutive equation for absolute permeability, and the most famous one is probably the Kozeny equation (also called Kozeny-Carman equation).

Dividing both sides of the equation by the area and using more general notation leads

$$q = -\frac{\kappa}{\mu}\nabla p,$$

where q is the flux (discharge per unit area, with units of length per time, m/s) and ∇p is the pressure gradient vector (Pa/m). This value of flux, often referred to as the Darcy flux or Darcy velocity, is not the velocity which the fluid traveling through the pores is experiencing. The fluid velocity (v) is related to the Darcy flux (q) by the porosity (φ). The flux is divided by porosity to account for the fact that only a fraction of the total formation volume is available for flow. The fluid velocity would be the velocity a conservative tracer would experience if carried by the fluid through the formation.

$$v = \frac{q}{\phi}.$$

Darcy's law is a simple mathematical statement which neatly summarizes several familiar properties that groundwater flowing in aquifers exhibits, including:

- if there is no pressure gradient over a distance, no flow occurs (these are hydrostatic conditions),
- if there is a pressure gradient, flow will occur from high pressure towards low pressure (opposite the direction of increasing gradient – hence the negative sign in Darcy's law),
- the greater the pressure gradient (through the same formation material), the greater the discharge rate, and
- the discharge rate of fluid will often be different – through different formation materials (or even through the same material, in a different direction) – even if the same pressure gradient exists in both cases.

A graphical illustration of the use of the steady-state groundwater flow equation (based on Darcy's law and the conservation of mass) is in the construction of flownets, to quantify the amount of groundwater flowing under a dam.

Darcy's law is only valid for slow, viscous flow; fortunately, most groundwater flow cases fall in this category. Typically any flow with a Reynolds number less than one is clearly laminar, and it would be valid to apply Darcy's law. Experimental tests have shown that flow regimes with Reynolds numbers up to 10 may still be Darcian, as in the case of groundwater flow. The Reynolds number (a dimensionless parameter) for porous media flow is typically expressed as

$$\mathrm{Re} = \frac{\rho v d_{30}}{\mu},$$

where ρ is the density of water (units of mass per volume), v is the specific discharge (not the pore velocity – with units of length per time), d_{30} is a representative grain diameter for the porous media (often taken as the 30% passing size from a grain size analysis using sieves – with units of length), and μ is the viscosity of the fluid.

Derivation

For stationary, creeping, incompressible flow, i.e. $\frac{D(\rho u_i)}{Dt} \approx 0,$ the Navier–Stokes equation simplifies to the Stokes equation:

$$\mu\nabla^2 u_i + \rho g_i - \partial_i p = 0,$$

where μ is the viscosity, u_i is the velocity in the i direction, g_i is the gravity component in the i direction and p is the pressure. Assuming the viscous resisting force is linear with the velocity we may write:

$$-\left(\kappa_{ij}\right)^{-1} \mu\phi u_j + \rho g_i - \partial_i p = 0,$$

where φ is the porosity, and κ_{ij} is the second order permeability tensor. This gives the velocity in the n direction,

$$\kappa_{ni}\left(\kappa_{ij}\right)^{-1} u_j = \delta_{nj} u_j = u_n = -\frac{\kappa_{ni}}{\phi\mu}\left(\partial_i p - \rho g_i\right),$$

which gives Darcy's law for the volumetric flux density in the n direction,

$$q_n = -\frac{\kappa_{ni}}{\mu}\left(\partial_i p - \rho g_i\right).$$

In isotropic porous media the off-diagonal elements in the permeability tensor are zero, $\kappa_{ij} = 0$ for $i \neq j$ and the diagonal elements are identical, $\kappa_{ii} = \kappa$, and the common form is obtained

$$\mathbf{q} = -\frac{\kappa}{\mu}\left(\nabla p - \rho \mathbf{g}\right).$$

The above equation is a governing equation for single phase fluid flow in a porous medium.

Additional Forms of Darcy's Law

Darcy's Law in Petroleum Engineering

Another derivation of Darcy's law is used extensively in petroleum engineering to determine the flow through permeable media — the most simple of which is for a one-dimensional, homogeneous rock formation with a single fluid phase and constant fluid viscosity.

$$Q = \frac{\kappa A}{\mu}\left(\frac{\partial p}{\partial x}\right),$$

where Q is the flowrate of the formation (in units of volume per unit time), k is the permeability of the formation (typically in millidarcys), A is the cross-sectional area of the formation, μ is the viscosity of the fluid (typically in units of centipoise). $\frac{\partial p}{\partial x}$ represents the pressure change per unit length of the formation. This equation can also be solved for permeability and is used to measure it, forcing a fluid of known viscosity through a core of a known length and area, and measuring the pressure drop across the length of the core.

Almost all oil reservoirs have a water zone blow the oil leg, and some have also a gas cap above the oil leg. When the reservoir pressure drops due to oil production, water flows into the oil zone from below, and gas flows into the oil zone from above (if the gas cap exists), and we get a simultaneous flow and immiscible mixing of all fluid phases in the oil zone. The operator of the oil field may also inject water (and/ or gas) in order to improve oil production. The petroleum industry is therefore using a generalized Darcy equation for multiphase flow that was developed by Muskat et alios. Because Darcy's name is so widespread and strongly associated with flow in porous media, the multiphase equation is denoted Darcy's law for multiphase flow or generalized Darcy equation (or law) or simply Darcy's equation (or law) or simply flow equation if the context says that the text is discussing the multiphase equation of Muskat et alios.

Darcy–Forchheimer Law

For flows in porous media with Reynolds numbers greater than about 1 to 10, inertial effects can also become significant. Sometimes an inertial term is added to the Darcy's equation, known as Forchheimer term. This term is able to account for the non-linear behavior of the pressure difference vs flow data.

$$\frac{\partial p}{\partial x} = -\frac{\mu}{\kappa} q - \frac{\rho}{\kappa_1} q^2,$$

where the additional term κ_1 is known as inertial permeability.

The flow in the middle of a sandstone reservoir is so slow that Forchheimer's equation is usually not needed, but the gas flow into a gas production well may be high enough to justify use of Forchheimer's equation. In this case the inflow performance calculations for the well, not the grid cell of the 3D model, is based on the Forchheimer equation. The effect of this is that an additional rate-dependent skin appears in the inflow performance formula.

Some carbonate reservoirs have lots of fractures, and Darcy's equation for multiphase flow is generalized in order to govern both flow in fractures and flow in the matrix (i.e. the traditional porous rock). The irregular surface of the fracture walls and high flow rate in the fractures, may justify use of Forchheimer's equation.

Darcy's Law for Gases in Fine Media (Knudsen Diffusion or Klinkenberg Effect)

For gas flow in small characteristic dimensions (e.g., very fine sand, nanoporous structures etc.), the particle-wall interactions become more frequent, giving rise to additional wall friction (Knudsen friction). For a flow in this region, where both viscous and Knudsen friction are present, a new formulation needs to be used. Knudsen presented a semi-empirical model for flow in transition regime based on his experiments on small capillaries. For a porous medium, the Knudsen equation can be given as

$$N = -\left(\frac{\kappa}{\mu}\frac{p_a + p_b}{2} + D_\mathrm{K}^\mathrm{eff}\right)\frac{1}{R_\mathrm{g}T}\frac{p_\mathrm{b} - p_\mathrm{a}}{L},$$

where N is the molar flux, R_g is the gas constant, T is the temperature, $D_\mathrm{K}^\mathrm{eff}$ is the effective Knudsen diffusivity of the porous media. The model can also be derived from the first-principle-based binary friction model (BFM). The differential equation of transition flow in porous media based on BFM is given as

$$\frac{\partial p}{\partial x} = -R_\mathrm{g}T\left(\frac{\kappa p}{\mu} + D_\mathrm{K}\right)^{-1} N.$$

This equation is valid for capillaries as well as porous media. The terminology of the Knudsen effect and Knudsen diffusivity is more common in mechanical and chemical engineering. In geological and petrochemical engineering, this effect is known as the Klinkenberg effect. Using the definition of molar flux, the above equation can be rewritten as

$$\frac{\partial p}{\partial x} = -R_\mathrm{g}T\left(\frac{\kappa p}{\mu} + D_\mathrm{K}\right)^{-1} \frac{p}{R_\mathrm{g}T} q.$$

This equation can be rearranged into the following equation

$$q = -\frac{\kappa}{\mu}\left(1 + \frac{D_\mathrm{K}\mu}{\kappa}\frac{1}{p}\right)\frac{\partial p}{\partial x}.$$

Comparing this equation with conventional Darcy's law, a new formulation can be given as

$$q = -\frac{\kappa^\mathrm{eff}}{\mu}\frac{\partial p}{\partial x},$$

where

$$\kappa^{\text{eff}} = \kappa\left(1 + \frac{D_{\text{K}}\mu}{\kappa}\frac{1}{p}\right).$$

This is equivalent to the effective permeability formulation proposed by Klinkenberg:

$$\kappa^{\text{eff}} = \kappa\left(1 + \frac{b}{p}\right).$$

where b is known as the Klinkenberg parameter, which depends on the gas and the porous medium structure. This is quite evident if we compare the above formulations. The Klinkenberg parameter b is dependent on permeability, Knudsen diffusivity and viscosity (i.e., both gas and porous medium properties).

Darcy's Law for Short Time Scales

For very short time scales, a time derivative of flux may be added to Darcy's law, which results in valid solutions at very small times (in heat transfer, this is called the modified form of Fourier's law),

$$\tau\frac{\partial q}{\partial t} + q = -\kappa\nabla h,$$

where τ is a very small time constant which causes this equation to reduce to the normal form of Darcy's law at "normal" times (> nanoseconds). The main reason for doing this is that the regular groundwater flow equation (diffusion equation) leads to singularities at constant head boundaries at very small times. This form is more mathematically rigorous, but leads to a hyperbolic groundwater flow equation, which is more difficult to solve and is only useful at very small times, typically out of the realm of practical use.

Brinkman form of Darcy's Law

Another extension to the traditional form of Darcy's law is the Brinkman term, which is used to account for transitional flow between boundaries (introduced by Brinkman in 1949),

$$\beta\nabla^2 q + q = -\frac{\kappa}{\mu}\nabla p,$$

where β is an effective viscosity term. This correction term accounts for flow through medium where the grains of the media are porous themselves, but is difficult to use, and is typically neglected.

Validity of Darcy's Law

Darcy's law is valid for laminar flow through sediments. In fine-grained sediments, the dimensions of interstices are small and thus flow is laminar. Coarse-grained sediments also behave similarly but in very coarse-grained sediments the flow may be turbulent. Hence Darcy's law is not always valid in such sediments. For flow through commercial pipes, the flow is laminar when Reynolds number is less than 2000, but in some sediments it has been found that flow is laminar when the value of Reynolds number is less than 1.

Navier–Stokes Equations

In physics, the Navier–Stokes equations named after Claude-Louis Navier and George Gabriel Stokes, describe the motion of viscous fluid substances. These balance equations arise from applying Newton's second law to fluid motion, together with the assumption that the stress in the fluid is the sum of a diffusing viscous term (proportional to the gradient of velocity) and a pressure term—hence describing *viscous flow*. The main difference between them and the simpler Euler equations for inviscid flow is that Navier–Stokes equations also factor in the Froude limit (no external field) and are not conservation equations, but rather a dissipative system, in the sense that they cannot be put into the quasilinear homogeneous form:

$$\mathbf{y}_t + \mathbf{A}(\mathbf{y})\mathbf{y}_x = 0.$$

Navier–Stokes equations are useful because they describe the physics of many phenomena of scientific and engineering interest. They may be used to model the weather, ocean currents, water flow in a pipe and air flow around a wing. The Navier–Stokes equations in their full and simplified forms help with the design of aircraft and cars, the study of blood flow, the design of power stations, the analysis of pollution, and many other things. Coupled with Maxwell's equations they can be used to model and study magnetohydrodynamics.

The Navier–Stokes equations are also of great interest in a purely mathematical sense. Somewhat surprisingly, given their wide range of practical uses, it has not yet been proven that in three dimensions solutions always exist, or that if they do exist, then they are smooth, i.e. they do not contain any mathematical singularity. These are called the Navier–Stokes existence and smoothness problems. The Clay Mathematics Institute has called this one of the seven most important open problems in mathematics and has offered a US$1,000,000 prize for a solution or a counterexample.

Flow Velocity

The solution of the Navier–Stokes equations is a flow velocity. It is a field, since it is defined at every point in a region of space and an interval of time. Once the velocity field is calculated other quantities of interest, such as pressure or temperature, may be found using additional equations and relations. This is different from what one normally sees in classical mechanics, where solutions are typically trajectories of position of a particle or deflection of a continuum. Studying velocity instead of position makes more sense for a fluid; however for visualization purposes one can compute various trajectories.

General Continuum Equations

The Navier–Stokes momentum equation can be derived as a particular form of the Cauchy momentum equation. In an inertial frame of reference, the conservation form of the equations of continuum motion is:

Cauchy momentum equation *(conservation form)*

$$\frac{\partial}{\partial t}(\rho \mathbf{u}) + \nabla \cdot (\rho \mathbf{u} \otimes \mathbf{u}) = -\nabla \cdot p\mathbf{I} + \nabla \cdot \boldsymbol{\tau} + \rho \mathbf{g}$$

where

ρ is the density,

u is the flow velocity,

∇ is the del operator.

p is the pressure

I is the identity matrix

τ is the deviatoric stress tensor, which has order two,

g represents body accelerations acting on the continuum, for example gravity, inertial accelerations, electrostatic accelerations, and so on.

$\otimes$ is the outer product: $u \otimes v = u \times v^T$

The left side of the equation describes acceleration, and may be composed of time-dependent and convective components (also the effects of non-inertial coordinates if present). The right side of the equation is in effect a summation of body forces (such as gravity), hydrostatic effects, and divergence of deviatoric stress.

In the Eulerian forms it is apparent that the assumption of an inviscid fluid, or no deviatoric stress, brings Cauchy equations to the Euler equations.

All non-relativistic balance equations, such as the Navier–Stokes equations, can be derived by beginning with the Cauchy equations and specifying the stress tensor through a constitutive relation. By expressing the shear tensor in terms of viscosity and fluid velocity, and assuming constant density and viscosity, the Cauchy equations will lead to the Navier–Stokes equations.

The incompressible case is simpler than the compressible one so for didactical purpose it should be presented before. However, the compressible case is the most general framework of Navier–Stokes equations so where not specified, Navier–Stokes equations are intended to be *compressible* Navier–Stokes equations.

Convective Acceleration

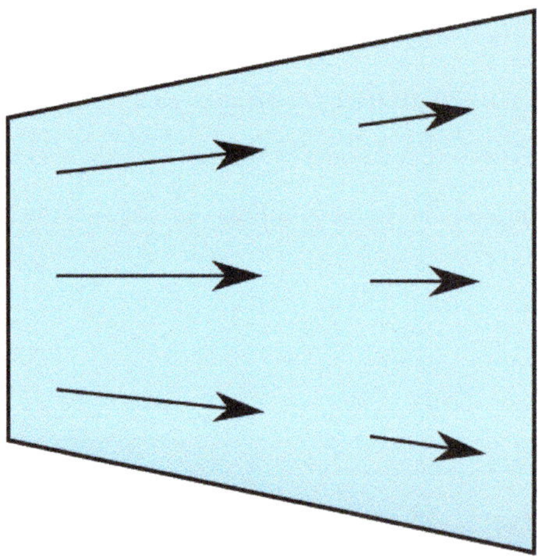

An example of convection. Though the flow may be steady (time-independent), the fluid decelerates as it moves down the diverging duct (assuming incompressible or subsonic compressible flow), hence there is an acceleration happening over position.

A significant feature of Cauchy equation and consequently all other continuum equations (including Euler and Navier–Stokes) is the presence of convective acceleration: the effect of acceleration of a flow with respect to space. While individual fluid particles indeed experience time-dependent acceleration, the convective acceleration of the flow field is a spatial effect, one example being fluid speeding up in a nozzle.

Incompressible Flow

The incompressible momentum Navier–Stokes equation results from the following assumptions on the Cauchy stress tensor:

- the stress is Galileian invariant: it does not depend directly on the flow velocity, but only on spatial derivatives of the flow velocity. So the stress variable is the tensor gradient ∇u.
- the fluid is assumed to be isotropic, as with gases and simple liquids, and consequently V is an isotropic tensor; furthermore, since the deviatoric stress tensor can be expressed in terms of the dynamic viscosity μ:

Stokes' stress constitutive equation *(expression used for incompressible elastic solids)*

$$\tau = 2\mu\varepsilon$$

where

$$\varepsilon = \frac{1}{2}\left(\nabla u + \nabla u^{\mathrm{T}}\right)$$

is the rate-of-strain tensor. So this decomposition can be explicited as:

Stokes's stress constitutive equation *(expression used for incompressible viscous fluids)*

$$\tau = \mu\left(\nabla u + \nabla u^{\mathrm{T}}\right)$$

Dynamic viscosity μ need not be constant – in incompressible flows it can depend on density and on pressure. Any equation expliciting one of these transport coefficient in the conservative variables is called an equation of state.

The divergence of the deviatoric stress is given by:

$$\nabla \cdot \tau = 2\mu\nabla \cdot \varepsilon = \mu\nabla \cdot \left(\nabla u + \nabla u^{\mathrm{T}}\right) = \mu\nabla^2 u$$

because $\nabla \cdot \mathbf{u} = 0$ for an incompressible fluid. Incompressibility rules out density and pressure waves like sound or shock waves, so this simplification is not useful if these phenomena are of interest. The incompressible flow assumption typically holds well with all fluids at low Mach numbers (say up to about Mach 0.3), such as for modelling air winds at normal temperatures. For incompressible (uniform density ρ0) flows the following identity holds:

$$\frac{1}{\rho_0}\nabla p = \nabla\left(\frac{p}{\rho_0}\right) \equiv \nabla w$$

where *w* is the specific (with the sense of *per unit mass*) thermodynamic work, the internal source term. Then the incompressible Navier–Stokes equations are best visualised by dividing for the density:

Incompressible Navier–Stokes equations (*convective form*)

$$\frac{\partial \mathbf{u}}{\partial t} + (\mathbf{u} \cdot \nabla)\mathbf{u} - \nu\nabla^2\mathbf{u} = -\nabla w + \mathbf{g}.$$

where $\nu = \mu/\rho_0$ is called the kinematic viscosity.

A laminar flow example

Velocity profile (laminar flow):

$$u(x)=0, \quad u(y)=u, \quad u(z)=0$$

for the y-direction, simplify the Navier–Stokes equation:

$$0=-\frac{dP}{dx}+\mu\left(\frac{d^2u}{dy^2}\right)$$

Integrate twice to find the velocity profile with boundary conditions $y = h$, $u = 0$, $y = -h$, $u = 0$:

$$u=\frac{1}{2\mu}\frac{dP}{dx}y^2+Ay+B$$

From this equation, substitute in the two boundary conditions to get two equations:

$$0=\frac{1}{2\mu}\frac{dP}{dx}h^2+Ah+B \qquad \frac{}{dx}h \quad Ah \quad B$$

$$0=\frac{1}{2\mu}\frac{dP}{dx}h^2-Ah+B$$

Add and solve for B:

$$B=-\frac{1}{2\mu}\frac{dP}{dx}h^2$$

Substitute and solve for A:

$$A=0$$

Finally this gives the velocity profile:

$$u=\frac{1}{2\mu}\frac{dP}{dx}\left(y^2-h^2\right)$$

in tensor notation:

Incompressible Navier–Stokes equations (*convective form*)

$$\left(\frac{\partial}{\partial t}+u_j\frac{\partial}{\partial x_j}-v\frac{\partial^2}{\partial x_j\partial x_j}\right)u_i=-\frac{\partial w}{\partial x_i}+g_i$$

It is well worth observing the meaning of each term (compare to the Cauchy momentum equation):

$$\overbrace{\underbrace{\frac{\partial \mathbf{u}}{\partial t}}_{\text{Variation}} + \underbrace{(\mathbf{u}\cdot\nabla)\mathbf{u}}_{\text{Convection}}}^{\text{Inertia (per volume)}} \overbrace{- \underbrace{\nu\nabla^2\mathbf{u}}_{\text{Diffusion}} = \underbrace{-\nabla w}_{\substack{\text{Internal}\\ \text{source}}}}^{\text{Divergence of stress}} + \underbrace{\mathbf{g}}_{\substack{\text{External}\\ \text{source}}}.$$

The higher-order term, namely the shear stress divergence $\nabla \cdot \tau$, has simply reduced to the vector laplacian term $\mu\nabla^2 u$. This laplacian term can be interpreted as the difference between the velocity at a point and the mean velocity in a small surrounding volume. This implies that – for a Newtonian fluid – viscosity operates as a *diffusion of momentum*, in much the same way as the heat conduction. In fact neglecting the convection term, incompressible Navier–Stokes equations lead to a vector diffusion equation (namely Stokes equations), but in general the convection term is present, so incompressible Navier–Stokes equations belong to the class of convection-diffusion equations.

In the usual case of an external field being a conservative field:

$$\mathbf{g} = -\nabla\phi$$

by defining the hydraulic head:

$$h \equiv w + \phi$$

one can finally condense the whole source in one term, arriving to the incompressible Navier–Stokes equation with conservative external field:

$$\frac{\partial \mathbf{u}}{\partial t} + (\mathbf{u}\cdot\nabla)\mathbf{u} - \nu\nabla^2\mathbf{u} = -\nabla h.$$

The incompressible Navier–Stokes equations with conservative external field is the fundamental equation of hydraulics. The domain for these equations is commonly a 3 or less Euclidean space, for which an orthogonal coordinate reference frame is usually set to explicit the system of scalar partial derivative equations to be solved. In 3D orthogonal coordinate systems are 3: Cartesian, cylindrical, and spherical. Expressing the Navier–Stokes vector equation in Cartesian coordinates is quite straightforward and not much influenced by the number of dimensions of the euclidean space employed, and this is the case also for the first-order terms (like the variation and convection ones) also in non-cartesian orthogonal coordinate systems. But for the higher order terms (the two coming from the divergence of the deviatoric stress that distinguish Navier–Stokes equations from Euler equations) some tensor calculus is required for deducing an expression in non-cartesian orthogonal coordinate systems.

The incompressible Navier–Stokes equation is composite, the sum of two orthogonal equations,

$$\frac{\partial \mathbf{u}}{\partial t} = \Pi^S \left(-(\mathbf{u} \cdot \nabla)\mathbf{u} + \nu \nabla^2 \mathbf{u} \right) + \mathbf{f}^S$$

$$\rho^{-1} \nabla p = \Pi^I \left(-(\mathbf{u} \cdot \nabla)\mathbf{u} + \nu \nabla^2 \mathbf{u} \right) + \mathbf{f}^I$$

where Π^S and Π^I are solenoidal and irrotational projection operators satisfying $\Pi^S + \Pi^I = 1$ and f^S and f^I are the non-conservative and conservative parts of the body force. This result follows from the Helmholtz Theorem (also known as the fundamental theorem of vector calculus). The first equation is a pressureless governing equation for the velocity, while the second equation for the pressure is a functional of the velocity and is related to the pressure Poisson equation.

The explicit functional form of the projection operator in 3D is found from the Helmholtz Theorem:

$$\Pi^S \mathbf{F}(\mathbf{r}) = \frac{1}{4\pi} \nabla \times \int \frac{\nabla' \times \mathbf{F}(\mathbf{r}')}{|\mathbf{r} - \mathbf{r}'|} dV', \quad \Pi^I = 1 - \Pi^S$$

with a similar structure in 2D. Thus the governing equation is an integro-differential equation similar to Coulomb and Biot-Savart law, not convenient for numerical computation.

An equivalent weak or variational form of the equation, proved to produce the same velocity solution as the Navier–Stokes equation, is given by,

$$\left(\mathbf{w}, \frac{\partial \mathbf{u}}{\partial t} \right) = -\left(\mathbf{w}, (\mathbf{u} \cdot \nabla)\mathbf{u} \right) - \nu \left(\nabla \mathbf{w} : \nabla \mathbf{u} \right) + \left(\mathbf{w}, \mathbf{f}^S \right)$$

for divergence-free test functions w satisfying appropriate boundary conditions. Here, the projections are accomplished by the orthogonality of the solenoidal and irrotational function spaces. The discrete form of this is eminently suited to finite element computation of divergence-free flow. There one will be able to address the question "How does one specify pressure-driven (Poiseuille) problems with a pressureless governing equation?".

The absence of pressure forces from the governing velocity equation demonstrates that the equation is not a dynamic one, but rather a kinematic equation where the divergence-free condition serves the role of a conservation equation. This all would seem to refute the frequent statements that the incompressible pressure enforces the divergence-free condition.

Discrete Velocity

With partitioning of the problem domain and defining basis functions on the partitioned domain, the discrete form of the governing equation is,

$$\left(\mathbf{w}_i, \frac{\partial \mathbf{u}_j}{\partial t}\right) = -\left(\mathbf{w}_i, (\mathbf{u}\cdot\nabla)\mathbf{u}_j\right) - \nu(\nabla\mathbf{w}_i : \nabla\mathbf{u}_j) + \left(\mathbf{w}_i, \mathbf{f}^S\right).$$

It is desirable to choose basis functions which reflect the essential feature of incompressible flow – the elements must be divergence-free. While the velocity is the variable of interest, the existence of the stream function or vector potential is necessary by the Helmholtz Theorem. Further, to determine fluid flow in the absence of a pressure gradient, one can specify the difference of stream function values across a 2D channel, or the line integral of the tangential component of the vector potential around the channel in 3D, the flow being given by Stokes' Theorem. Discussion will be restricted to 2D in the following.

We further restrict discussion to continuous Hermite finite elements which have at least first-derivative degrees-of-freedom. With this, one can draw a large number of candidate triangular and rectangular elements from the plate-bending literature. These elements have derivatives as components of the gradient. In 2D, the gradient and curl of a scalar are clearly orthogonal, given by the expressions,

$$\nabla\phi = \left(\frac{\partial\phi}{\partial x}, \frac{\partial\phi}{\partial y}\right)^{\mathrm{T}},$$

$$\nabla\times\phi = \left(\frac{\partial\phi}{\partial y}, -\frac{\partial\phi}{\partial x}\right)^{\mathrm{T}}.$$

Adopting continuous plate-bending elements, interchanging the derivative degrees-of-freedom and changing the sign of the appropriate one gives many families of stream function elements.

Taking the curl of the scalar stream function elements gives divergence-free velocity elements. The requirement that the stream function elements be continuous assures that the normal component of the velocity is continuous across element interfaces, all that is necessary for vanishing divergence on these interfaces.

Boundary conditions are simple to apply. The stream function is constant on no-flow surfaces, with no-slip velocity conditions on surfaces. Stream function differences across open channels determine the flow. No boundary conditions are necessary on open boundaries, though consistent values may be used with some problems. These are all Dirichlet conditions.

The algebraic equations to be solved are simple to set up, but of course are non-linear, requiring iteration of the linearized equations.

Similar considerations apply to three-dimensions, but extension from 2D is not immediate because of the vector nature of the potential, and there exists no simple relation between the gradient and the curl as was the case in 2D.

Pressure Recovery

Recovering pressure from the velocity field is easy. The discrete weak equation for the pressure gradient is,

$$(\mathbf{g}_i, \nabla p) = -\left(\mathbf{g}_i, (\mathbf{u}\cdot\nabla)\mathbf{u}_j\right) - \nu\left(\nabla\mathbf{g}_i : \nabla\mathbf{u}_j\right) + \left(\mathbf{g}_i, \mathbf{f}^I\right)$$

where the test/weight functions are irrotational. Any conforming scalar finite element may be used. However, the pressure gradient field may also be of interest. In this case one can use scalar Hermite elements for the pressure. For the test/weight functions g_i one would choose the irrotational vector elements obtained from the gradient of the pressure element.

Compressible Flow

The Navier–Stokes equations result from the following assumptions on the stress tensor:

- the stress is Galileian invariant: it does not depend directly on the flow velocity, but only on spatial derivatives of the flow velocity. So the stress variable is the tensor gradient ∇u.
- the stress is linear in this variable: $\sigma(\nabla u) = C : (\nabla u)$, where C is the fourth-order tensor representing the constant of proportionality, called the viscosity or elasticity tensor, and : is the double-dot product.
- the fluid is assumed to be isotropic, as with gases and simple liquids, and consequently V is an isotropic tensor; furthermore, since the stress tensor is symmetric, by Helmholtz decomposition it can be expressed in terms of two scalar Lamé parameters, the bulk viscosity λ and the dynamic viscosity μ, as it is usual in linear elasticity:

Linear stress constitutive equation *(expression used for elastic solid)*

$$\sigma = \lambda(\nabla\cdot u)I + 2\mu\varepsilon$$

where I is the identity tensor, $\varepsilon(\nabla u) \equiv 1/2\nabla u + 1/2(\nabla u)^T$ is the rate-of-strain tensor and $\nabla \cdot u$ is the rate of expansion of the flow. So this decomposition can be explicited as:

$$\sigma = \lambda(\nabla\cdot u)I + \mu\left(\nabla u + (\nabla u)^{\mathrm{T}}\right).$$

Since the trace of the rate-of-strain tensor in three dimensions is:

$$\mathrm{tr}(\varepsilon) = \nabla\cdot u.$$

The trace of the stress tensor in three dimensions becomes:

$$\mathrm{tr}(\sigma) = (3\lambda + 2\mu)\nabla \cdot u.$$

So by alternatively decomposing the stress tensor is into isotropic and deviatoric parts, as usual in fluid dynamics:

$$\sigma = \left(\lambda + \tfrac{2}{3}\mu\right)(\nabla \cdot u)I + \mu\left(\nabla u + (\nabla u)^{\mathrm{T}} - \tfrac{2}{3}(\nabla \cdot u)I\right)$$

Introducing the second viscosity ζ,

$$\zeta \equiv \lambda + \tfrac{2}{3}\mu,$$

we arrive to the linear constitutive equation in the form usually employed in thermal hydraulics:

Linear stress constitutive equation *(expression used for fluids)*

$$\sigma = \zeta(\nabla \cdot u)I + \mu\left(\nabla u + (\nabla u)^{\mathrm{T}} - \tfrac{2}{3}(\nabla \cdot u)I\right)$$

Both bulk viscosity ζ and dynamic viscosity μ need not be constant – in general, they depend on density, on each other (the viscosity is expressed in pressure), and in compressible flows also on temperature. Any equation expliciting one of these transport coefficient in the conservation variables is called an equation of state.

By computing the divergence of the stress tensor, since the divergence of tensor ∇u is $\nabla^2 u$ and the divergence of tensor $(\nabla u)^{\mathrm{T}}$ is $\nabla(\nabla \cdot u)$, one finally arrives to the compressible (most general) Navier–Stokes momentum equation:

Navier–Stokes momentum equation (*convective form*)

$$\frac{\partial \mathbf{u}}{\partial t} + \mathbf{u} \cdot \nabla \mathbf{u} = -\frac{1}{\rho}\nabla \bar{p} + \nu\nabla^2 \mathbf{u} + \tfrac{1}{3}\nu\nabla(\nabla \cdot \mathbf{u}) + \mathbf{g}.$$

Bulk viscosity is assumed to be constant, otherwise it should not be taken out of the last derivative. The effect of the volume viscosity ζ is that the mechanical pressure is not equivalent to the thermodynamic pressure:

$$\bar{p} \equiv p - \zeta\nabla \cdot \mathbf{u},$$

This difference is usually neglected, sometimes by explicitly assuming $\zeta = 0$, but it could have an impact in sound absorption and attenuation and shock waves.

For the special case of an incompressible flow, the pressure constrains the flow so that the volume of fluid elements is constant: isochoric flow resulting in a solenoidal velocity field with $\nabla \cdot u = 0$.

Other Equations

The Navier–Stokes equations are strictly a statement of the balance of momentum. To fully describe fluid flow, more information is needed, how much depending on the assumptions made. This additional information may include boundary data (no-slip, capillary surface, etc.), conservation of mass, balance of energy, and/or an equation of state.

Continuity Equation

Regardless of the flow assumptions, a statement of the conservation of mass is generally necessary. This is achieved through the mass continuity equation, given in its most general form as:

$$\frac{\partial \rho}{\partial t} + \nabla \cdot (\rho \mathbf{u}) = 0$$

or, using the substantive derivative:

$$\frac{D\rho}{Dt} + \rho(\nabla \cdot \mathbf{u}) = 0.$$

In the example below we can assume to have a Newtonian fluid as well as having ρ and μ both be constant.

Recall that mass continuity is simply the summation of the rate of mass in and the rate of mass out.

rate of mass accumulated = rate of mass in – rate of mass out

$$\Delta x \Delta y \Delta z \frac{\partial \rho}{\partial t} = \Delta y \Delta z \left(\rho U_x |_x - \rho U_x |_{x+\Delta x} \right) + \Delta x \Delta z \left(\rho U_y |_y - \rho U_y |_{y+\Delta y} \right) + \Delta x \Delta y \left(\rho U_z |_z - \rho U_z |_{z+\Delta z} \right)$$

$$\Rightarrow \frac{\partial \rho}{\partial t} = \frac{\rho U_x |_x - \rho U_x |_{x+\Delta x}}{\Delta x} + \frac{\rho U_y |_y - \rho U_y |_{y+\Delta y}}{\Delta y} + \frac{\rho U_z |_z - \rho U_z |_{z+\Delta z}}{\Delta z}$$

$$= \rho \left(\frac{U_x |_x - U_x |_{x+\Delta x}}{\Delta x} + \frac{U_y |_y - U_y |_{y+\Delta y}}{\Delta y} + \frac{U_z |_z - U_z |_{z+\Delta z}}{\Delta z} \right)$$

$$= -\rho \left(\frac{\partial U_x}{\partial x} + \frac{\partial U_y}{\partial y} + \frac{\partial U_z}{\partial z} \right)$$

Since there is no change in density ρ over time, $\partial \rho / \partial t = 0$, we have:

$$0 = -\rho\left(\frac{\partial U_x}{\partial x} + \frac{\partial U_y}{\partial y} + \frac{\partial U_z}{\partial z}\right)$$

Recall that ρ is a constant thus proving the divergence theorem above.

Stream Function for 2D Equations

Taking the curl of the Navier–Stokes equation results in the elimination of pressure. This is especially easy to see if 2D Cartesian flow is assumed (like in the degenerate 3D case with $u_z = 0$ and no dependence of anything on z), where the equations reduce to:

$$\rho\left(\frac{\partial u_x}{\partial t} + u_x\frac{\partial u_x}{\partial x} + u_y\frac{\partial u_x}{\partial y}\right) = -\frac{\partial p}{\partial x} + \mu\left(\frac{\partial^2 u_x}{\partial x^2} + \frac{\partial^2 u_x}{\partial y^2}\right) + \rho g_x$$

$$\rho\left(\frac{\partial u_y}{\partial t} + u_x\frac{\partial u_y}{\partial x} + u_y\frac{\partial u_y}{\partial y}\right) = -\frac{\partial p}{\partial y} + \mu\left(\frac{\partial^2 u_y}{\partial x^2} + \frac{\partial^2 u_y}{\partial y^2}\right) + \rho g_y.$$

Differentiating the first with respect to y, the second with respect to x and subtracting the resulting equations will eliminate pressure and any conservative force. Defining the stream function ψ through

$$u_x = \frac{\partial \psi}{\partial y}; \quad u_y = -\frac{\partial \psi}{\partial x}$$

results in mass continuity being unconditionally satisfied (given the stream function is continuous), and then incompressible Newtonian 2D momentum and mass conservation condense into one equation:

$$\frac{\partial}{\partial t}\left(\nabla^2\psi\right) + \frac{\partial \psi}{\partial y}\frac{\partial}{\partial x}\left(\nabla^2\psi\right) - \frac{\partial \psi}{\partial x}\frac{\partial}{\partial y}\left(\nabla^2\psi\right) = \nu\nabla^4\psi$$

where ∇^4 is the 2D biharmonic operator and ν is the kinematic viscosity, $\nu = \mu/\rho$. We can also express this compactly using the Jacobian determinant:

$$\frac{\partial}{\partial t}\left(\nabla^2\psi\right) + \frac{\partial\left(\psi, \nabla^2\psi\right)}{\partial(y, x)} = \nu\nabla^4\psi.$$

This single equation together with appropriate boundary conditions describes 2D fluid flow, taking only kinematic viscosity as a parameter. Note that the equation for creeping flow results when the left side is assumed zero.

In axisymmetric flow another stream function formulation, called the Stokes stream function, can be used to describe the velocity components of an incompressible flow with one scalar function.

The incompressible Navier–Stokes equation is a differential algebraic equation, having the inconvenient feature that there is no explicit mechanism for advancing the pressure in time. Consequently, much effort has been expended to eliminate the pressure from all or part of the computational process. The stream function formulation eliminates the pressure but only in two dimensions and at the expense of introducing higher derivatives and elimination of the velocity, which is the primary variable of interest.

Properties

Nonlinearity

The Navier–Stokes equations are nonlinear partial differential equations in the general case and so remain in almost every real situation. In some cases, such as one-dimensional flow and Stokes flow (or creeping flow), the equations can be simplified to linear equations. The nonlinearity makes most problems difficult or impossible to solve and is the main contributor to the turbulence that the equations model.

The nonlinearity is due to convective acceleration, which is an acceleration associated with the change in velocity over position. Hence, any convective flow, whether turbulent or not, will involve nonlinearity. An example of convective but laminar (nonturbulent) flow would be the passage of a viscous fluid (for example, oil) through a small converging nozzle. Such flows, whether exactly solvable or not, can often be thoroughly studied and understood.

Turbulence

Turbulence is the time-dependent chaotic behavior seen in many fluid flows. It is generally believed that it is due to the inertia of the fluid as a whole: the culmination of time-dependent and convective acceleration; hence flows where inertial effects are small tend to be laminar (the Reynolds number quantifies how much the flow is affected by inertia). It is believed, though not known with certainty, that the Navier–Stokes equations describe turbulence properly.

The numerical solution of the Navier–Stokes equations for turbulent flow is extremely difficult, and due to the significantly different mixing-length scales that are involved in turbulent flow, the stable solution of this requires such a fine mesh resolution that the computational time becomes significantly infeasible for calculation or direct numerical simulation. Attempts to solve turbulent flow using a laminar solver typically result in a time-unsteady solution, which fails to converge appropriately. To counter this, time-averaged equations such as the Reynolds-averaged Navier–Stokes equations (RANS), supplemented with turbulence models, are used in practical computational fluid dynamics (CFD) applications when modeling turbulent flows. Some models include the Spalart–Allmaras, k–ω, k–ε, and SST models, which add a variety of additional equations to bring closure to the RANS equations. Large eddy simulation (LES) can also be used to solve these equations numerically. This approach is computationally more expensive—in time and in computer memory—than RANS, but produces better results because it explicitly resolves the larger turbulent scales.

Applicability

Together with supplemental equations (for example, conservation of mass) and well formulated boundary conditions, the Navier–Stokes equations seem to model fluid motion accurately; even turbulent flows seem (on average) to agree with real world observations.

The Navier–Stokes equations assume that the fluid being studied is a continuum (it is infinitely divisible and not composed of particles such as atoms or molecules), and is not moving at relativistic velocities. At very small scales or under extreme conditions, real fluids made out of discrete molecules will produce results different from the continuous fluids modeled by the Navier–Stokes equations. For example, capillarity of internal layers in fluids appears for flow with high gradients. For large Knudsen number of the problem, the Boltzmann equation may be a suitable replacement. Failing that, one may have to resort to molecular dynamics or various hybrid methods.

Another limitation is simply the complicated nature of the equations. Time-tested formulations exist for common fluid families, but the application of the Navier–Stokes equations to less common families tends to result in very complicated formulations and often to open research problems. For this reason, these equations are usually rewritten for Newtonian fluids where the viscosity model is linear; truly general models for the flow of other kinds of fluids (such as blood) do not exist.

Application to Specific Problems

The Navier–Stokes equations, even when written explicitly for specific fluids, are rather generic in nature and their proper application to specific problems can be very diverse. This is partly because there is an enormous variety of problems that may be modeled, ranging from as simple as the distribution of static pressure to as complicated as multiphase flow driven by surface tension.

Generally, application to specific problems begins with some flow assumptions and initial/boundary condition formulation, this may be followed by scale analysis to further simplify the problem.

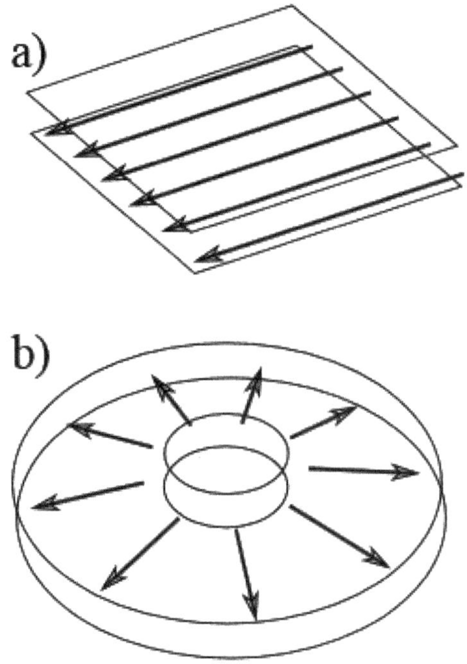

Visualization of (a) parallel flow and (b) radial flow.

Parallel Flow

Assume steady, parallel, one dimensional, non-convective pressure-driven flow between parallel plates, the resulting scaled (dimensionless) boundary value problem is:

$$\frac{d^2u}{dy^2} = -1; \quad u(0) = u(1) = 0.$$

The boundary condition is the no slip condition. This problem is easily solved for the flow field:

$$u(y) = \frac{y - y^2}{2}.$$

From this point onward more quantities of interest can be easily obtained, such as viscous drag force or net flow rate.

Radial Flow

Difficulties may arise when the problem becomes slightly more complicated. A seemingly modest twist on the parallel flow above would be the *radial* flow between parallel plates; this involves convection and thus non-linearity. The velocity field may be represented by a function $f(z)$ that must satisfy:

$$\frac{d^2 f}{dz^2} + Rf^2 = -1; \quad f(-1) = f(1) = 0.$$

This ordinary differential equation is what is obtained when the Navier–Stokes equations are written and the flow assumptions applied (additionally, the pressure gradient is solved for). The nonlinear term makes this a very difficult problem to solve analytically (a lengthy implicit solution may be found which involves elliptic integrals and roots of cubic polynomials). Issues with the actual existence of solutions arise for $R > 1.41$ (approximately; this is not $\sqrt{2}$), the parameter R being the Reynolds number with appropriately chosen scales. This is an example of flow assumptions losing their applicability, and an example of the difficulty in "high" Reynolds number flows.

Exact Solutions of the Navier–Stokes Equations

Some exact solutions to the Navier–Stokes equations exist. Examples of degenerate cases — with the non-linear terms in the Navier–Stokes equations equal to zero — are Poiseuille flow, Couette flow and the oscillatory Stokes boundary layer. But also more interesting examples, solutions to the full non-linear equations, exist; for example the Taylor–Green vortex. Note that the existence of these exact solutions does not imply they are stable: turbulence may develop at higher Reynolds numbers.

Under additional assumptions, the component parts can be separated.

A two dimensional example

For example, in the case of an unbounded planar domain with two-dimensional — incompressible and stationary — flow in polar coordinates (r,φ), the velocity components (u_r,u_φ) and pressure p are:

$$u_r = \frac{A}{r},$$

$$u_\phi = B\left(\frac{1}{r} - r^{\frac{A}{\nu}+1}\right),$$

$$p = -\frac{A^2 + B^2}{2r^2} - \frac{2B^2 \nu r^{\frac{A}{\nu}}}{A} + \frac{B^2 r^{\left(\frac{2A}{\nu}+2\right)}}{\frac{2A}{\nu}+2}$$

where A and B are arbitrary constants. This solution is valid in the domain $r \geq 1$ and for $A < -2\nu$.

In Cartesian coordinates, when the viscosity is zero ($\nu = 0$), this is:

$$\mathbf{v}(x, y) = \frac{1}{x^2 + y^2}\begin{pmatrix} Ax + By \\ Ay - Bx \end{pmatrix},$$

$$p(x,y) = -\frac{A^2+B^2}{2\left(x^2+y^2\right)}$$

A three-dimensional example

For example, in the case of an unbounded Euclidean domain with three-dimensional — incompressible, stationary and with zero viscosity (ν = 0) — radial flow in Cartesian coordinates (x,y,z), the velocity vector v and pressure p are:

$$\mathbf{v}(x,y,z) = \frac{A}{x^2+y^2+z^2}\begin{pmatrix} x \\ y \\ z \end{pmatrix},$$

$$p(x,y,z) = -\frac{A^2}{2\left(x^2+y^2+z^2\right)}.$$

There is a singularity at $x = y = z = 0$.

A Three-dimensional Steady-state Vortex Solution

Some of the flow lines along a Hopf fibration.

A steady-state example with no singularities comes from considering the flow along the lines of a Hopf fibration. Let r be a constant radius of the inner coil. One set of solutions is given by:

$$\rho(x,y,z) = \frac{3B}{r^2+x^2+y^2+z^2}\, p(x,y,z) = \frac{-A^2B}{\left(r^2+x^2+y^2+z^2\right)^3}\, \mathbf{u}(x,y,z) =$$

$$\frac{A}{\left(r^2+x^2+y^2+z^2\right)^2}\begin{pmatrix} 2(-ry+xz) \\ 2(rx+yz) \\ r^2-x^2-y^2+z^2 \end{pmatrix} g = 0 \mu \ = 0$$

for arbitrary constants A and B. This is a solution in a non-viscous gas (compressible fluid) whose density, velocities and pressure goes to zero far from the origin. (Note this is not a solution to the Clay Millennium problem because that refers to incompressible fluids where ρ is a constant, neither does it deal with the uniqueness of the Navier–Stokes equations with respect to any turbulence properties.) It is also worth pointing out that the components of the velocity vector are exactly those from the Pythagorean quadruple parametrization. Other choices of density and pressure are possible with the same velocity field:

Other Choices of Density and Pressure

Another choice of pressure and density with the same velocity vector above is one where the pressure and density fall to zero at the origin and are highest in the central loop at $z = 0, x^2 + y^2 = r^2$:

$$\rho(x,y,z)=\frac{20B\left(x^2+y^2\right)}{\left(r^2+x^2+y^2+z^2\right)^3}\; p(x,y,z)=\frac{-A^2B}{\left(r^2+x^2+y^2+z^2\right)^4}+\frac{-4A^2B\left(x^2+y^2\right)}{\left(r^2+x^2+y^2+z^2\right)^5}.$$

In fact in general there are simple solutions for any polynomial function f where the density is:

$$\rho(x,y,z)=\frac{1}{r^2+x^2+y^2+z^2}f\left(\frac{x^2+y^2}{\left(r^2+x^2+y^2+z^2\right)^2}\right).$$

Wyld Diagrams

Wyld diagrams are bookkeeping graphs that correspond to the Navier–Stokes equations via a perturbation expansion of the fundamental continuum mechanics. Similar to the Feynman diagrams in quantum field theory, these diagrams are an extension of Keldysh's technique for nonequilibrium processes in fluid dynamics. In other words, these diagrams assign graphs to the (often) turbulent phenomena in turbulent fluids by allowing correlated and interacting fluid particles to obey stochastic processes associated to pseudo-random functions in probability distributions.

Representations in 3D

Cartesian Coordinates

From the general form of the Navier–Stokes, with the velocity vector expanded as $u = (u_x, u_y, u_z)$, sometimes respectively named u, v, w, we may write the vector equation explicitly,

$$x:\rho\left(\frac{\partial u_x}{\partial t}+u_x\frac{\partial u_x}{\partial x}+u_y\frac{\partial u_x}{\partial y}+u_z\frac{\partial u_x}{\partial z}\right)=-\frac{\partial p}{\partial x}+\mu\left(\frac{\partial^2 u_x}{\partial x^2}+\frac{\partial^2 u_x}{\partial y^2}+\frac{\partial^2 u_x}{\partial z^2}\right)-\mu\frac{\partial}{\partial x}\left(\frac{\partial u_x}{\partial x}+\frac{\partial u_y}{\partial y}+\frac{\partial u_z}{\partial z}\right)+\rho g_x$$

$$y:\rho\left(\frac{\partial u_y}{\partial t}+u_x\frac{\partial u_y}{\partial x}+u_y\frac{\partial u_y}{\partial y}+u_z\frac{\partial u_y}{\partial z}\right)=-\frac{\partial p}{\partial y}+\mu\left(\frac{\partial^2 u_y}{\partial x^2}+\frac{\partial^2 u_y}{\partial y^2}+\frac{\partial^2 u_y}{\partial z^2}\right)-\mu\frac{\partial}{\partial y}\left(\frac{\partial u_x}{\partial x}+\frac{\partial u_y}{\partial y}+\frac{\partial u_z}{\partial z}\right)+\rho g_y$$

$$z:\rho\left(\frac{\partial u_z}{\partial t}+u_x\frac{\partial u_z}{\partial x}+u_y\frac{\partial u_z}{\partial y}+u_z\frac{\partial u_z}{\partial z}\right)=-\frac{\partial p}{\partial z}+\mu\left(\frac{\partial^2 u_z}{\partial x^2}+\frac{\partial^2 u_z}{\partial y^2}+\frac{\partial^2 u_z}{\partial z^2}\right)-\mu\frac{\partial}{\partial z}\left(\frac{\partial u_x}{\partial x}+\frac{\partial u_y}{\partial y}+\frac{\partial u_z}{\partial z}\right)+\rho g_z$$

Note that gravity has been accounted for as a body force, and the values of g_x, g_y, g_z will depend on the orientation of gravity with respect to the chosen set of coordinates.

The continuity equation reads:

$$\frac{\partial \rho}{\partial t}+\frac{\partial(\rho u_x)}{\partial x}+\frac{\partial(\rho u_y)}{\partial y}+\frac{\partial(\rho u_z)}{\partial z}=0.$$

When the flow is incompressible, ρ does not change for any fluid particle, and its material derivative vanishes: $D\rho/Dt = 0$. The continuity equation is reduced to:

$$\frac{\partial u_x}{\partial x}+\frac{\partial u_y}{\partial y}+\frac{\partial u_z}{\partial z}=0.$$

Thus, for the incompressible version of the Navier–Stokes equation the second part of the viscous terms fall away.

This system of four equations comprises the most commonly used and studied form. Though comparatively more compact than other representations, this is still a nonlinear system of partial differential equations for which solutions are difficult to obtain.

Cylindrical coordinates

A change of variables on the Cartesian equations will yield the following momentum equations for r, φ, and z

$$r:\rho\left(\frac{\partial u_r}{\partial t}+u_r\frac{\partial u_r}{\partial r}+\frac{u_\phi}{r}\frac{\partial u_r}{\partial \phi}+u_z\frac{\partial u_r}{\partial z}-\frac{u_\phi^2}{r}\right)=-\frac{\partial p}{\partial r}+\mu\left(\frac{1}{r}\frac{\partial}{\partial r}\left(r\frac{\partial u_r}{\partial r}\right)+\frac{1}{r^2}\frac{\partial^2 u_r}{\partial \phi^2}+\frac{\partial^2 u_r}{\partial z^2}-\frac{u_r}{r^2}-\frac{2}{r^2}\frac{\partial u_\phi}{\partial \phi}\right)+\rho g_r$$

$$\phi:\rho\left(\frac{\partial u_\phi}{\partial t}+u_r\frac{\partial u_\phi}{\partial r}+\frac{u_\phi}{r}\frac{\partial u_\phi}{\partial \phi}+u_z\frac{\partial u_\phi}{\partial z}+\frac{u_r u_\phi}{r}\right)=-\frac{1}{r}\frac{\partial p}{\partial \phi}+\mu\left(\frac{1}{r}\frac{\partial}{\partial r}\left(r\frac{\partial u_\phi}{\partial r}\right)+\frac{1}{r^2}\frac{\partial^2 u_\phi}{\partial \phi^2}+\frac{\partial^2 u_\phi}{\partial z^2}+\frac{2}{r^2}\frac{\partial u_r}{\partial \phi}-\frac{u_\phi}{r^2}\right)+\rho g_\phi$$

$$z:\rho\left(\frac{\partial u_z}{\partial t}+u_r\frac{\partial u_z}{\partial r}+\frac{u_\phi}{r}\frac{\partial u_z}{\partial \phi}+u_z\frac{\partial u_z}{\partial z}\right)=-\frac{\partial p}{\partial z}+\mu\left(\frac{1}{r}\frac{\partial}{\partial r}\left(r\frac{\partial u_z}{\partial r}\right)+\frac{1}{r^2}\frac{\partial^2 u_z}{\partial \phi^2}+\frac{\partial^2 u_z}{\partial z^2}\right)+\rho g_z$$

The gravity components will generally not be constants, however for most applications either the coordinates are chosen so that the gravity components are constant or else it is assumed that gravity is counteracted by a pressure field (for example, flow in horizontal pipe is treated normally without gravity and without a vertical pressure gradient). The continuity equation is:

$$\frac{\partial \rho}{\partial t}+\frac{1}{r}\frac{\partial}{\partial r}(\rho r u_r)+\frac{1}{r}\frac{\partial(\rho u_\phi)}{\partial \phi}+\frac{\partial(\rho u_z)}{\partial z}=0.$$

This cylindrical representation of the incompressible Navier–Stokes equations is the second most commonly seen (the first being Cartesian above). Cylindrical coordinates are chosen to take advantage of symmetry, so that a velocity component can disappear. A very common case is axisymmetric flow with the assumption of no tangential velocity ($u_\varphi = 0$), and the remaining quantities are independent of φ:

$$\rho\left(\frac{\partial u_r}{\partial t}+u_r\frac{\partial u_r}{\partial r}+u_z\frac{\partial u_r}{\partial z}\right)=-\frac{\partial p}{\partial r}+\mu\left(\frac{1}{r}\frac{\partial}{\partial r}\left(r\frac{\partial u_r}{\partial r}\right)+\frac{\partial^2 u_r}{\partial z^2}-\frac{u_r}{r^2}\right)+\rho g_r$$

$$\rho\left(\frac{\partial u_z}{\partial t}+u_r\frac{\partial u_z}{\partial r}+u_z\frac{\partial u_z}{\partial z}\right)=-\frac{\partial p}{\partial z}+\mu\left(\frac{1}{r}\frac{\partial}{\partial r}\left(r\frac{\partial u_z}{\partial r}\right)+\frac{\partial^2 u_z}{\partial z^2}\right)+\rho g_z$$

$$\frac{1}{r}\frac{\partial}{\partial r}(r u_r)+\frac{\partial u_z}{\partial z}=0.$$

Navier–Stokes Equations use in Games

The Navier–Stokes equations are used extensively in video games in order to model a wide variety of natural phenomena. Simulations of small-scale gaseous fluids, such as fire and smoke, are often based on the seminal paper "Real-Time Fluid Dynamics for Games" by Jos Stam, which elaborates one of the methods proposed in Stam's earlier, more famous paper "Stable Fluids" from 1999. Stam proposes stable fluid simulation using a Navier–Stokes solution method from 1968, coupled with an unconditionally stable semi-Lagrangian advection scheme, as first proposed in 1992.

More recent implementations based upon this work run on the game systems graphics processing unit (GPU) as opposed to the central processing unit (CPU) and achieve a much higher degree of performance. Many improvements have been proposed to Stam's original work, which suffers inherently from high numerical dissipation in both velocity and mass.

An introduction to interactive fluid simulation can be found in the 2007 ACM SIGGRAPH course, Fluid Simulation for Computer Animation.

Permeability of Different Soils

Permeability (k) is an engineering property of soils and is a function of the soil type. Its value depends on the average size of the pores and is related to the distribution of particle sizes, particle shape and soil structure. The ratio of permeabilities of typical sands/

gravels to those of typical clays is of the order of 10^6. A small proportion of fine material in a coarse-grained soil can lead to a significant reduction in permeability.

For different soil types as per grain size, the orders of magnitude for permeability are as follows:

Soil	**k (cm/sec)**
Gravel	10^0
Coarse sand	10^0 to 10^{-1}
Medium sand	10^{-1} to 10^{-2}
Fine sand	10^{-2} to 10^{-3}
Silty sand	10^{-3} to 10^{-4}
Silt	1×10^{-5}
Clay	10^{-7} to 10^{-9}

Factors Affecting Permeability

In soils, the permeant or pore fluid is mostly water whose variation in property is generally very less. Permeability of all soils is strongly influenced by the density of packing of the soil particles, which can be represented by void ratio (e) or porosity (n).

For Sands

In sands, permeability can be empirically related to the square of some representative grain size from its grain-size distribution. For filter sands, Allen Hazen in 1911 found that $k \approx 100\,(D_{10})^2$ cm/s where D_{10} = effective grain size in cm.

Different relationships have been attempted relating void ratio and permeability, such as $k \propto e^3/(1+e)$, and $k \propto e^2$. They have been obtained from the Kozeny-Carman equation for laminar flow in saturated soils.

$$k=\frac{1}{k_0 k_T S_S^2}\cdot\frac{e^3}{1+e}\cdot\frac{\gamma_w}{\eta}$$

where k_o and k_T are factors depending on the shape and tortuosity of the pores respectively, S_S is the surface area of the solid particles per unit volume of solid material, and γ_w and η are unit weight and viscosity of the pore water. The equation can be reduced to a simpler form as

$$k=C.\frac{e^3}{1+e}\approx C.e^2$$

For Silts and Clays

For silts and clays, the Kozeny-Carman equation does not work well, and log k versus e plot has been found to indicate a linear relationship.

For clays, it is typically found that

$$\log_{10} k = \frac{e - e_k}{C_k}$$

where C_k is the permeability change index and e_k is a reference void ratio.

References

- "Chuquicamata's Life Underground Will Cost a Fortune, but is Likely to Pay Off for Codelco | Copper Investing News". 2015-04-06. Archived from the original on April 6, 2015. Retrieved 2015-06-11
- "The Spanish Tradition in Gold and Silver Mining." Otis E. Young Arizona and the West, Vol. 7, No. 4 (Winter, 1965), pp. 299-314 (Journal of the Southwest)
- Chesworth, Ward (2008). Encyclopedia of soil science. Dordrecht, Netherlands: Springer. p. 694. ISBN 1402039948. Retrieved 2 July 2016
- Swaziland Natural Trust Commission, "Cultural Resources - Malolotja Archaeology, Lion Cavern," Retrieved August 27, 2007, [1]
- Cambell, Bonnie (2008). "Regulation & Legitimacy in the Mining Industry in Africa: Where does" (PDF). Review of African Political Economy. 35 (3): 367–389. doi:10.1080/03056240802410984. Retrieved 7 April 2011
- Soil Science Glossary Terms Committee (2008). Glossary of Soil Science Terms 2008. Madison, WI: Soil Science Society of America. ISBN 978-0-89118-851-3
- Peace Parks Foundation, "Major Features: Cultural Importance." Republic of South Africa: Author. Retrieved August 27, 2007, [2]
- Acheson, D. J. (1990), Elementary Fluid Dynamics, Oxford Applied Mathematics and Computing Science Series, Oxford University Press, ISBN 0-19-859679-0
- "Assessing the opportunities of landfill mining - Research database - University of Groningen". www.rug.nl. Retrieved 2015-06-11
- Landau, L. D.; Lifshitz, E. M. (1987), Fluid mechanics, Course of Theoretical Physics, 6 (2nd revised ed.), Pergamon Press, ISBN 0-08-033932-8, OCLC 15017127
- "The Independent, 20 Jan. 2007: ''The end of a Celtic tradition: the last gold miner in Wales''". News.independent.co.uk. 2007-01-20. Archived from the original on July 6, 2008. Retrieved 2013-06-22
- Polyanin, A. D.; Kutepov, A. M.; Vyazmin, A. V.; Kazenin, D. A. (2002), Hydrodynamics, Mass and Heat Transfer in Chemical Engineering, Taylor & Francis, London, ISBN 0-415-27237-8
- The Cerrillos Hills Park Coalition, (2000). Cerrillos Hills Historic Park Vision Statement. Public documents: Author. Retrieved August 27, 2007, [4]. Archived August 1, 2012, at the Wayback Machine

5

Soil Conservation: An Integrated Study

Over usage, erosion and acidification causes loss of fertility in land. Soil conservation involves methods of avoiding erosion and exhaustion. This section helps the readers in developing a better understanding of soil erosion. Soil mechanics is best understood in confluence with the major topics listed in the following chapter.

Soil Conservation

Soil conservation is the preventing of soil loss from erosion or reduced fertility caused by over usage, [Soil acidification acidification] [salinization] or other chemical soil contamination.

Slash-and-burn and other unsustainable methods of subsistence farming are practiced in some lesser developed areas. A sequel to the deforestation is typically large scale erosion, loss of soil nutrients and sometimes total desertification.

Erosion barriers on disturbed slope, Marin County, California

Techniques for improved soil conservation include crop rotation, cover crops, conservation tillage and planted windbreaks and affect both erosion and fertility. When plants, especially trees, die, they decay and become part of the soil. Code 330 defines standard methods recommended by the US Natural Resources Conservation Service.

Contour plowing, Pennsylvania 1938. The rows formed slow water run-off during rainstorms to prevent soil erosion and allows the water time to infiltrate into the soil.

Farmers have practiced soil conservation for millennia. Conservation practices fall in multiple categories:

Contour Plowing

Contour plowing orients crop furrows following the contour lines of the farmed area. Furrows move left and right to maintain a constant altitude, which reduces runoff. Contour plowing was practiced by the ancient Phoenicians, and is effective for slopes between two and ten percent. Contour plowing can increase crop yields from 10 to 50 percent, partially as a result of greater soil retention.

Terracing or Terrace Farming

Terracing is the practice of creating nearly level areas in a hillside area. The terraces form a series of steps, each at a higher level than the previous. Terraces are protected from erosion by other soil barriers.

Terraced farming is more common on small farms and in underdeveloped countries, since mechanized equipment is difficult to deploy in this setting. It protects the soil from its erosion. It is one of the way by which soil erosion can be stopped.

Keyline Design

Keyline design is an enhancement of contour farming, where the total watershed properties are taken into account in forming the contour lines.

Perimeter Runoff Control

Tree, shrubs and ground-cover are effective perimeter treatment for soil erosion prevention, by impeding surface flows. A special form of this perimeter or inter-row treatment is the use of a "grass way" that both channels and dissipates runoff through surface friction, impeding surface runoff and encouraging infiltration of the slowed surface water.

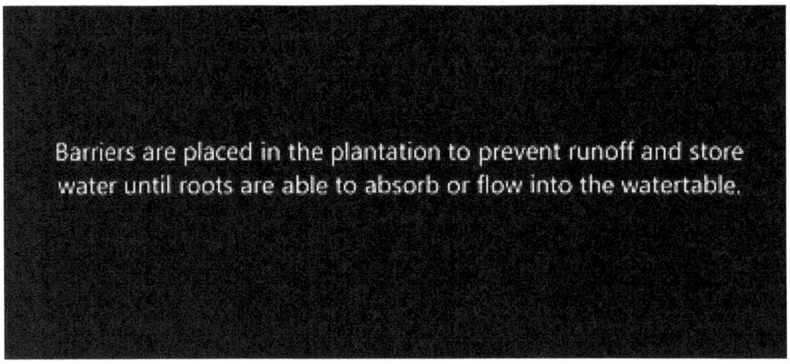

runoff and filter soxx

Windbreaks

Windbreaks are sufficiently dense rows of trees at the windward exposure of an agricultural field subject to wind erosion. Evergreen species provide year-round protection; however, as long as foliage is present in the seasons of bare soil surfaces, the effect of deciduous trees may be adequate.

Cover Crops/crop Rotation

Cover crops such as legumes, white turnip, radishes and other species are rotated with cash crops to blanket the soil year-round and act as green manure that replenishes nitrogen and other critical nutrients. Cover crops also help suppress weeds.

Soil-conservation Farming

Soil-conservation farming involves no-till farming, "green manures" and other soil-enhancing practices. Such farming methods attempt to mimic the biology of virgin land. They can revive damaged soil, minimize erosion, encourage plant growth, eliminate the use of nitrogen fertilizer or fungicide, produce above-average yields and protect crops during droughts or flooding. The result is less labor and lower costs that increase farmers' profits. No-till farming and cover crops act as sinks for nitrogen and other nutrients. This increases the amount of soil organic matter.

Repeated plowing/tilling degrades soil, killing its beneficial fungi and earthworms. Once damaged, soil may take multiple seasons to fully recover, even in optimal circumstances.

Critics argue that no-till and related methods are impractical and too expensive for many growers, partly because it requires new equipment. They cite advantages for conventional tilling depending on the geography, crops and soil conditions. Some farmers claimed that no-till complicates weed control, delays planting and that post-harvest residues, especially for corn, are hard to manage.

Salinity Management

Salt deposits on the former bed of the Aral Sea

Salinity in soil is caused by irrigating with salty water. Water then evaporates from the soil leaving the salt behind. Salt breaks down the soil structure, causing infertility and reduced growth.

The ions responsible for salination are: sodium (Na^+), potassium (K^+), calcium (Ca^{2+}), magnesium (Mg^{2+}) and chlorine (Cl^-). Salinity is estimated to affect about one third of the earth's arable land. Soil salinity adversely affects crop metabolism and erosion usually follows.

Salinity occurs on drylands from overirrigation and in areas with shallow saline water tables. Over-irrigation deposits salts in upper soil layers as a byproduct of soil infiltration; irrigation merely increases the rate of salt deposition. The best-known case of shallow saline water table capillary action occurred in Egypt after the 1970 construction of the Aswan Dam. The change in the groundwater level led to high salt concentrations in the water table. The continuous high level of the water table led to soil salination.

Use of humic acids may prevent excess salination, especially given excessive irrigation. Humic acids can fix both anions and cations and eliminate them from root zones. Planting species that can tolerate saline conditions to produce surface cover can preserve soil salinity can be reduced. Salt-tolerant plants include saltbush, a plant found in much of North America and in the Mediterranean regions of Europe.

Soil Organisms

When worms excrete egesta in the form of casts, a balanced selection of minerals and plant nutrients is made into a form accessible for root uptake. Earthworm casts are five times richer in available nitrogen, seven times richer in available phosphates and eleven times richer in available potash than the surrounding upper 150 millimetres (5.9 in) of soil. The weight of casts produced may be greater than 4.5 kg per worm per year. By burrowing, the earthworm improves soil porosity, creating channels that enhance the processes of aeration and drainage.

Yellow fungus, a mushroom that assists in organic decay.

Other important soil organisms include nematodes, mycorrizha and bacteria.

Degraded soil requires synthetic fertilizer to produce high yields. Lacking structure increases erosion and carries nitrogen and other pollutants into rivers and streams.

Each one percent increase in soil organic matter helps soil hold 20,000 gallons more water per acre.

Mineralization

To allow plants full realization of their phytonutrient potential, active mineralization of the soil is sometimes undertaken. This can involve adding crushed rock or chemical soil supplements. In either case the purpose is to combat mineral depletion. A broad range of minerals can be used, including common substances such as phosphorus and more exotic substances such as zinc and selenium. Extensive research examines the phase transitions of minerals in soil with aqueous contact.

Flooding can bring significant sediment to an alluvial plain. While this effect may not be desirable if floods endanger life or if the sediment originates from productive land, this process of addition to a floodplain is a natural process that can rejuvenate soil chemistry through mineralization.

Soil Formation

The term "soil" can have different meanings, depending upon the field in which it is considered.

To a geologist, it is the material in the relative thin zone of the Earth's surface within which roots occur, and which are formed as the products of past surface processes. The rest of the crust is grouped under the term "rock".

To a pedologist, it is the substance existing on the surface, which supports plant life.

To an engineer, it is a material that can be:

- built on: foundations of buildings, bridges
- built in: basements, culverts, tunnels
- built with: embankments, roads, dams
- supported: retaining walls

Soil Mechanics is a discipline of Civil Engineering involving the study of soil, its behaviour and application as an engineering material.

Soil Mechanics is the application of laws of mechanics and hydraulics to engineering problems dealing with sediments and other unconsolidated accumulations of solid particles, which are produced by the mechanical and chemical disintegration of rocks, regardless of whether or not they contain an admixture of organic constituents.

Soil consists of a multiphase aggregation of solid particles, water, and air. This fundamental composition gives rise to unique engineering properties, and the description of its mechanical behavior requires some of the most classic principles of engineering mechanics.

Engineers are concerned with soil's mechanical properties: permeability, stiffness, and strength. These depend primarily on the nature of the soil grains, the current stress, the water content and unit weight.

Soil

Soil is a mixture of minerals, organic matter, gases, liquids, and countless organisms that together support life on Earth. Soil is a natural body called the pedosphere which has four important functions: it is a medium for plant growth; it is a means of water storage, supply and purification; it is a modifier of Earth's atmosphere; it is a habitat for organisms; all of which, in turn, modify the soil.

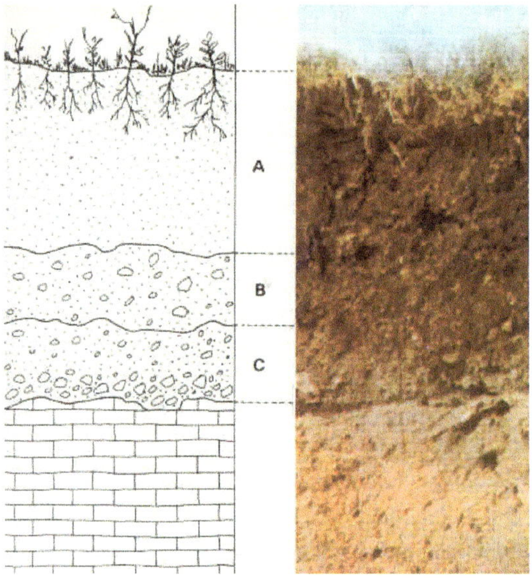

A, B, and C represent the soil profile, a notation firstly coined by Vasily Dokuchaev, the father of pedology; A is the topsoil; B is a regolith; C is a saprolite, a less-weathered regolith; the bottom-most layer represents the bedrock.

Surface-water-gley developed in glacial till, Northern Ireland.

Soil has been called the *Skin of the Earth* and interfaces with the lithosphere, the hydrosphere, the atmosphere, and the biosphere. The term *pedolith*, used commonly to refer to the soil, literally translates *ground stone*. Soil consists of a solid phase of minerals (the soil matrix) and organic matter, as well as a porous phase that holds gases (the soil atmosphere) and water (the soil solution). Accordingly, soils are often treated as a three-state system of solids, liquids, and gases.

Soil is a product of the influence of climate, relief (elevation, orientation, and slope of terrain), organisms, and its parent materials (original minerals) interacting over time. Soil continually undergoes development by way of numerous physical, chemical and biological processes, which include weathering with associated erosion. Given its complexity and strong internal connectedness soil has been considered as an ecosystem by soil ecologists.

Most soils have a dry bulk density (density of soil taking into account voids when dry) between 1.1 and 1.6 g/cm^3, while the soil particle density is much higher, in the range of 2.6 to 2.7 g/cm^3. Little of the soil of planet Earth is older than the Pleistocene and none is older than the Cenozoic, although fossilized soils are preserved from as far back as the Archean.

Soil science has two basic branches of study: edaphology and pedology. Edaphology is concerned with the influence of soils on living things. Pedology is focused on the formation, description (morphology), and classification of soils in their natural environment. In engineering terms, soil is referred to as regolith, or loose rock material that lies above the *solid geology*. Soil is commonly referred to as *earth* or *dirt*; technically, the term *dirt* should be restricted to displaced soil.

As soil resources serve as a basis for food security, the international community advocates its sustainable and responsible use through different types of soil governance.

Overview

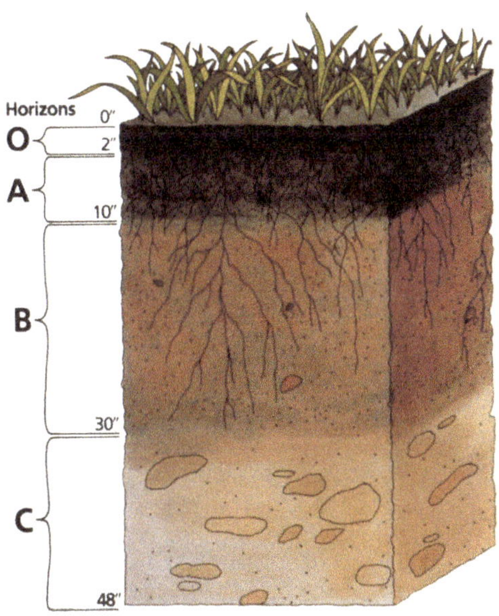

Soil Profile: Darkened topsoil and reddish subsoil layers are typical in some regions.

Soil is a major component of the Earth's ecosystem. The world's ecosystems are impacted in far-reaching ways by the processes carried out in the soil, from ozone depletion and global warming, to rainforest destruction and water pollution. Following the atmosphere, the soil is the next largest carbon reservoir on Earth, and it is potentially one of the most reactive to human disturbance and climate change. As the planet warms, soils will add carbon dioxide to the atmosphere due to its increased biological activity at higher temperatures. Thus, soil carbon losses likely have a large positive feedback response (amplification) to global warming, although positive feedback has been questioned on the base of more recent knowledge on soil carbon turnover.

Soil acts as an engineering medium, a habitat for soil organisms, a recycling system for nutrients and organic wastes, a regulator of water quality, a modifier of atmospheric composition, and a medium for plant growth, in other terms one of the world's best providers of ecosystem services. Since soil has a tremendous range of available niches and habitats, it contains most of the Earth's genetic diversity. A gram of soil can contain billions of organisms, belonging to thousands of species, mostly microbial and in the main still unexplored. Soil has a mean prokaryotic density of roughly 10^8 organisms per gram, whereas the ocean has no more than 10^7 procaryotic organisms per milliliter (gram) of seawater. The carbon content of the soil is eventually returned to the atmosphere through the process of respiration carried out by heterotrophic organisms that feed upon the carbonaceous material in the soil, but a substantial part is retained in the soil in the form of humus, the ratio respired to fixed carbon decreasing with cultivation. Since plant roots need oxygen, ventilation is an important characteristic of soil. This ventilation can be accomplished via networks of interconnected soil pores, which also absorb and hold rainwater making it readily available for plant uptake. Since plants require a nearly continuous supply of water, but most regions receive sporadic rainfall, the water-holding capacity of soils is vital for plant survival.

Soils can effectively remove impurities, kill disease agents, and degrade contaminants, this latter property being called natural attenuation. Typically, soils maintain a net absorption of oxygen and methane, and undergo a net release of carbon dioxide and nitrous oxide. Soils offer plants physical support, air, water, temperature moderation, nutrients, and protection from toxins. Soils provide readily available nutrients to plants and animals by converting dead organic matter into various nutrient forms.

Description

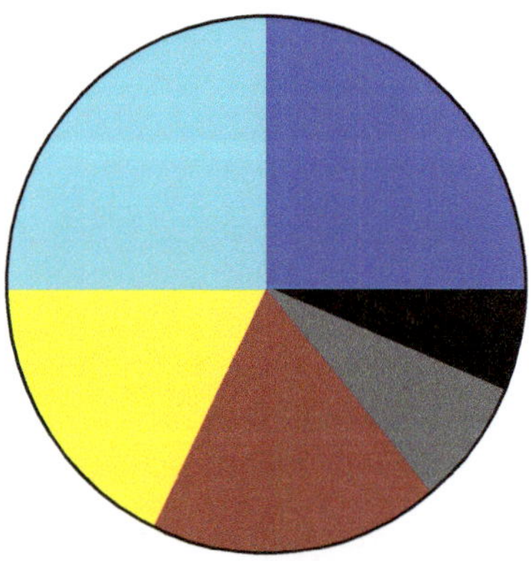

Components of a loam soil by percent volume

Water (25%)

Gases (25%)

Sand (18%)

Silt (18%)

Clay (9%)

Organic matter (5%)

Soils supply plants with nutrients that are held in place by their clay and humus content. For optimum plant growth, the soil components by volume should be roughly 50% solids (45% mineral and 5% organic matter), and 50% voids of which half is occupied by water and half by gas. The percent soil mineral and organic content is typically treated as a constant, while the percent soil water and gas content is considered highly variable whereby a rise in one is simultaneously balanced by a reduction in the other. The pore space allows for the infiltration and movement of air and water, both of which are critical for life in soil. Compaction, a common problem with soils, reduces this space, preventing air and water from reaching the plant roots and soil organisms.

Given sufficient time, an undifferentiated soil will evolve a soil profile which consists of two or more layers, referred to as soil horizons, that differ in one or more properties such as in their texture, structure, density, porosity, consistency, temperature, color, and reactivity. The horizons differ greatly in thickness and generally lack sharp boundaries. Soil profile development is dependent on the processes that form soils from their parent materials, the type of parent material, and the factors that control soil formation. The biological influences on soil properties are strongest near the surface, while the geochemical influences on soil properties increase with depth. Mature soil profiles in temperate climate regions typically include three basic master horizons: A, B and C. The solum normally includes the A and B horizons. The living component of the soil is largely confined to the solum. In the more hot, humid, climate of the tropics, a soil may have only a single horizon, when all the rock material has been converted to soil (residual soil).

The soil texture is determined by the relative proportions of sand, silt, and clay in the soil. The addition of organic matter, water, gases and time-related biotic and abiotic processes causes that texture to develop into a larger soil structure called an aggregate or a ped. At that point a soil can be said to be developed, and can be described further in terms of color, porosity, consistency, reaction (acidity), etc.

Of all the factors influencing the evolution of soil, water is the most powerful due to its involvement in the solution, erosion, transportation, and deposition of the materials of which a soil is composed. The mixture of water and dissolved or suspended materials that occupy the soil pore space is called the soil solution. Since soil water is never pure water, but contains hundreds of dissolved organic and mineral substances, it may be more accurately called the soil solution. Water is central to the solution, precipitation and leaching of minerals from the soil profile. Finally, water affects the type of vegetation that grows in a soil, which in turn affects the development of the soil, a complex feedback which is exemplified in the dynamics of banded vegetation patterns in semi-arid regions.

The most influential factors in stabilizing soil fertility are the soil colloids, clay and humus. Soil colloids behave as repositories of nutrients and moisture and so act to buffer the variations of soil solution ions and moisture. The contribution of soil colloids to soil nutrition is out of proportion to their part of the soil. Colloids act to store nutrients that might otherwise be leached from the soil or to release those ions in response to changes of soil pH, and so, make them available to plants.

The greatest influence on plant nutrient availability is soil pH, which is a measure of the hydrogen ion (acid-forming) soil reactivity, and is in turn a function of the soil materials, precipitation level, and plant root behavior. Soil pH strongly affects the availability of nutrients.

Most nutrients, with the exception of nitrogen, originate from minerals. Some nitrogen originates from rain as dilute nitric acid and ammonia, but most of the nitrogen is available in soils as a result of nitrogen fixation by bacteria. The action of microbes on organic matter and minerals may be to free nutrients for use, sequester them, or cause their loss from the soil by their volatilisation to gases or their leaching from the soil. The nutrients may be stored on soil colloids, or within live or dead organic matter, but they may not be accessible to plants due to extremes of pH.

The organic material of the soil has a powerful effect on its development, fertility, and available moisture. Following water and soil colloids, organic material is next in importance to a soil's formation and fertility.

History of the Study of Soil

Studies Concerning Soil Fertility

The history of the study of soil is intimately tied to our urgent need to provide food for ourselves and forage for our animals. Throughout history, civilizations have prospered or declined as a function of the availability and productivity of their soils.

The Greek historian Xenophon (450–355 B.C.) is credited with being the first to expound upon the merits of green-manuring crops: "But then whatever weeds are upon the ground, being turned into earth, enrich the soil as much as dung."

Columella's "Husbandry," circa 60 A.D., advocated the use of lime and that clover and alfalfa (green manure) should be turned under, and was used by 15 generations (450 years) under the Roman Empire until its collapse. From the fall of Rome to the French Revolution, knowledge of soil and agriculture was passed on from parent to child and as a result, crop yields were low. During the European Dark Ages, Yahya Ibn al-'Awwam's handbook, with its emphasis on irrigation, guided the people of North Africa, Spain and the Middle East; a translation of this work was finally carried to the southwest of the United States when under Spanish influence. Olivier de Serres, considered as the father of French agronomy, was the first to suggest the abandonment of fallowing and its replacement by hay meadows within crop rotations, and he highlighted the importance of soil (the French terroir) in the management of vineyards. His famous book *Le Théâtre d'Agriculture et mesnage des champs* contributed to the rise of modern, sustainable agriculture and to the collapse of old agricultural practices such as the lifting of forest litter for the amendment of crops (the French *soutrage*) and assarting, which ruined the soils of western Europe during Middle Ages and even later on according to regions.

Experiments into what made plants grow first led to the idea that the ash left behind when plant matter was burned was the essential element but overlooked the role of nitrogen, which is not left on the ground after combustion, a belief which prevailed until the 19th century. In about 1635, the Flemish chemist Jan Baptist van Helmont thought he had proved water to be the essential element from his famous five years' experiment with a willow tree grown with only the addition of rainwater. His conclusion came from the fact that the increase in the plant's weight had apparently been produced only by the addition of water, with no reduction in the soil's weight. John Woodward (d. 1728) experimented with various types of water ranging from clean to muddy and found muddy water the best, and so he concluded that earthy matter was the essential element. Others concluded it was humus in the soil that passed some essence to the growing plant. Still others held that the vital growth principal was something passed from dead plants or animals to the new plants. At the start of the 18th century, Jethro Tull demonstrated that it was beneficial to cultivate (stir) the soil, but his opinion that the stirring made the fine parts of soil available for plant absorption was erroneous.

As chemistry developed, it was applied to the investigation of soil fertility. The French chemist Antoine Lavoisier showed in about 1778 that plants and animals must [combust] oxygen internally to live and was able to deduce that most of the 165-pound weight of van Helmont's willow tree derived from air. It was the French agriculturalist Jean-Baptiste Boussingault who by means of experimentation obtained evidence showing that the main sources of carbon, hydrogen and oxygen for plants were air and water, while nitrogen was taken from soil. Justus von Liebig in his book *Organic chemistry in its applications to agriculture and physiology* (published 1840), asserted that the chemicals in plants must have come from the soil and air and that to maintain soil fertility, the used minerals must be replaced. Liebig nevertheless believed the nitrogen was supplied from the air. The enrichment of soil with guano by the Incas was rediscovered in 1802, by Alexander von Humboldt. This led to its mining and that of Chilean nitrate and to its application to soil in the United States and Europe after 1840.

The work of Liebig was a revolution for agriculture, and so other investigators started experimentation based on it. In England John Bennet Lawes and Joseph Henry Gilbert worked in the Rothamsted Experimental Station, founded by the former, and (re)discovered that plants took nitrogen from the soil, and that salts needed to be in an available state to be absorbed by plants. Their investigations also produced the "superphosphate", consisting in the acid treatment of phosphate rock. This led to the invention and use of salts of potassium (K) and nitrogen (N) as fertilizers. Ammonia generated by the production of coke was recovered and used as fertiliser. Finally, the chemical basis of nutrients delivered to the soil in manure was understood and in the mid-19th century chemical fertilisers were applied. However, the dynamic interaction of soil and its life forms still awaited discovery.

In 1856 J. Thomas Way discovered that ammonia contained in fertilisers was transformed into nitrates, and twenty years later Robert Warington proved that this transformation was done by living organisms. In 1890 Sergei Winogradsky announced he had found the bacteria responsible for this transformation.

It was known that certain legumes could take up nitrogen from the air and fix it to the soil but it took the development of bacteriology towards the end of the 19th century to lead to an understanding of the role played in nitrogen fixation by bacteria. The symbiosis of bacteria and leguminous roots, and the fixation of nitrogen by the bacteria, were simultaneously discovered by German agronomist Hermann Hellriegel and Dutch microbiologist Martinus Beijerinck.

Crop rotation, mechanisation, chemical and natural fertilisers led to a doubling of wheat yields in Western Europe between 1800 and 1900.

Studies Concerning Soil Formation

The scientists who studied the soil in connection with agricultural practices had considered it mainly as a static substrate. However, soil is the result of evolution from more ancient geological materials. After studies of the improvement of the soil commenced, others began to study soil genesis and as a result also soil types and classifications.

In 1860, in Mississippi, Eugene W. Hilgard studied the relationship among rock material, climate, and vegetation, and the type of soils that were developed. He realised that the soils were dynamic, and considered soil types classification. Unfortunately his work was not continued. At the same time Vasily Dokuchaev (about 1870) was leading a team of soil scientists in Russia who conducted an extensive survey of soils, finding that similar basic rocks, climate and vegetation types lead to similar soil layering and types, and established the concepts for soil classifications. Due to the language barriers, the work of this team was not communicated to Western Europe until 1914 by a publication in German by K. D. Glinka, a member of the Russian team.

Curtis F. Marbut was influenced by the work of the Russian team, translated Glinka's publication into English, and as he was placed in charge of the U. S. National Cooperative Soil Survey, applied it to a national soil classification system.

Soil-forming Processes

Soil formation, or pedogenesis, is the combined effect of physical, chemical, biological and anthropogenic processes working on soil parent material. Soil is said to be formed when organic matter has accumulated and colloids are washed downward, leaving deposits of clay, humus, iron oxide, carbonate, and gypsum, producing a distinct layer called the B horizon. This is a somewhat arbitrary definition as mixtures of sand, silt, clay and humus will support biological and agricultural activity before that time. These constituents are moved from one level to another by water and animal activity. As a result, layers (horizons) form in the soil profile. The alteration and movement of materials within a soil causes the formation of distinctive soil horizons.

How soil formation proceeds is influenced by at least five classic factors that are intertwined in the evolution of a soil. They are: parent material, climate, topography (relief), organisms, and time. When reordered to climate, relief, organisms, parent material, and time, they form the acronym CROPT.

An example of the development of a soil would begin with the weathering of lava flow bedrock, which would produce the purely mineral-based parent material from which the soil texture forms. Soil development would proceed most rapidly from bare rock of recent flows in a warm climate, under heavy and frequent rainfall. Under such conditions, plants become established very quickly on basaltic lava, even though there is very little organic material. The plants are supported by the porous rock as it is filled with nutrient-bearing water that carries dissolved minerals from the rocks and guano. Crevasses and pockets, local topography of the rocks, would hold fine materials and harbour plant roots. The developing plant roots are associated with mycorrhizal fungi that assist in breaking up the porous lava, and by these means organic matter and a finer mineral soil accumulate with time.

Parent Material

The mineral material from which a soil forms is called parent material. Rock, whether its origin is igneous, sedimentary, or metamorphic, is the source of all soil mineral materials and the origin of all plant nutrients with the exceptions of nitrogen, hydrogen and carbon. As the parent material is chemically and physically weathered, transported, deposited and precipitated, it is transformed into a soil.

Typical soil parent mineral materials are:

- Quartz: SiO_2

- Calcite: $CaCO_3$
- Feldspar: $KAlSi_3O_8$
- Mica (biotite): $K(Mg,Fe)_3AlSi_3O_{10}(OH)_2$

Classification of Parent Material

Loess field in Germany.

Parent materials are classified according to how they came to be deposited. Residual materials are mineral materials that have weathered in place from primary bedrock. Transported materials are those that have been deposited by water, wind, ice or gravity. Cumulose material is organic matter that has grown and accumulates in place.

Residual soils are soils that develop from their underlying parent rocks and have the same general chemistry as those rocks. The soils found on mesas, plateaux, and plains are residual soils. In the United States as little as three percent of the soils are residual.

Most soils derive from transported materials that have been moved many miles by wind, water, ice and gravity.

- Aeolian processes (movement by wind) are capable of moving silt and fine sand many hundreds of miles, forming loess soils (60–90 percent silt), common in the Midwest of North America and in Central Asia. Clay is seldom moved by wind as it forms stable aggregates.
- Water-transported materials are classed as either alluvial, lacustrine, or marine. Alluvial materials are those moved and deposited by flowing water. Sedimentary deposits settled in lakes are called lacustrine. Lake Bonneville and many soils around the Great Lakes of the United States are examples. Marine deposits, such as soils along the Atlantic and Gulf Coasts and in the Imperial Valley of California of the United States, are the beds of ancient seas that have been revealed as the land uplifted.

- Ice moves parent material and makes deposits in the form of terminal and lateral moraines in the case of stationary glaciers. Retreating glaciers leave smoother ground moraines and in all cases, outwash plains are left as alluvial deposits are moved downstream from the glacier.

- Parent material moved by gravity is obvious at the base of steep slopes as talus cones and is called colluvial material.

Cumulose parent material is not moved but originates from deposited organic material. This includes peat and muck soils and results from preservation of plant residues by the low oxygen content of a high water table. While peat may form sterile soils, muck soils may be very fertile.

Weathering of Parent Material

The weathering of parent material takes the form of physical weathering (disintegration), chemical weathering (decomposition) and chemical transformation. Generally, minerals that are formed under the high temperatures and pressures at great depths within the Earth's mantle are less resistant to weathering, while minerals formed at low temperature and pressure environment of the surface are more resistant to weathering. Weathering is usually confined to the top few meters of geologic material, because physical, chemical, and biological stresses generally decrease with depth. Physical disintegration begins as rocks that have solidified deep in the Earth are exposed to lower pressure near the surface and swell and become mechanically unstable. Chemical decomposition is a function of mineral solubility, the rate of which doubles with each 10 °C rise in temperature, but is strongly dependent on water to effect chemical changes. Rocks that will decompose in a few years in tropical climates will remain unaltered for millennia in deserts. Structural changes are the result of hydration, oxidation, and reduction.

- Physical disintegration is the first stage in the transformation of parent material into soil. Temperature fluctuations cause expansion and contraction of the rock, splitting it along lines of weakness. Water may then enter the cracks and freeze and cause the physical splitting of material along a path toward the center of the rock, while temperature gradients within the rock can cause exfoliation of "shells". Cycles of wetting and drying cause soil particles to be abraded to a finer size, as does the physical rubbing of material as it is moved by wind, water, and gravity. Water can deposit within rocks minerals that expand upon drying, thereby stressing the rock. Finally, organisms reduce parent material in size through the action of plant roots or digging on the part of animals.

- Chemical decomposition and structural changes result when minerals are made soluble by water or are changed in structure. The first three of the following list are solubility changes and the last three are structural changes.

1. The solution of salts in water results from the action of bipolar water on ionic salt compounds producing a solution of ions and water, removing those minerals and reducing the rock's integrity.

2. Hydrolysis is the transformation of minerals into polar molecules by the splitting of intervening water. This results in soluble acid-base pairs. For example, the hydrolysis of orthoclase-feldspar transforms it to acid silicate clay and basic potassium hydroxide, both of which are more soluble.

3. In carbonation, the solution of carbon dioxide in water forms carbonic acid. Carbonic acid will transform calcite into more soluble calcium bicarbonate.

4. Hydration is the inclusion of water in a mineral structure, causing it to swell and leaving it more stressed and easily decomposed.

5. Oxidation of a mineral compound is the inclusion of oxygen in a mineral, causing it to increase its oxidation number and swell due to the relatively large size of oxygen, leaving it stressed and more easily attacked by water (hydrolysis) or carbonic acid (carbonation).

6. Reduction the opposite of oxidation, means the removal of oxygen, hence the oxidation number of some part of the mineral is reduced, which occurs when oxygen is scarce. The reduction of minerals leaves them electrically unstable, more soluble and internally stressed and easily decomposed.

Of the above, hydrolysis and carbonation are the most effective.

Saprolite is a particular example of a residual soil formed from the transformation of granite, metamorphic and other types of bedrock into clay minerals. Often called "weathered granite", saprolite is the result of weathering processes that include: hydrolysis, chelation from organic compounds, hydration (the solution of minerals in water with resulting cation and anion pairs) and physical processes that include freezing and thawing. The mineralogical and chemical composition of the primary bedrock material, its physical features, including grain size and degree of consolidation, and the rate and type of weathering transforms the parent material into a different mineral. The texture, pH and mineral constituents of saprolite are inherited from its parent material.

Climate

The principal climatic variables influencing soil formation are effective precipitation (i.e., precipitation minus evapotranspiration) and temperature, both of which affect the rates of chemical, physical, and biological processes. The temperature and moisture both influence the organic matter content of soil through their effects on the balance between plant growth and microbial decomposition. Climate is the dominant factor in soil formation, and soils show the distinctive characteristics of the climate zones in which they form. For every 10 °C rise in temperature, the rates of biochemical reactions more than double. Mineral precipitation and temperature are the primary climatic influences on soil formation. If warm temperatures and abundant water are present in the profile at the same time, the processes of weathering, leaching, and plant growth will be maximized. Humid climates favor the growth of trees. In contrast, grasses are the dominant native vegetation in subhumid and semiarid regions, while shrubs and brush of various kinds dominate in arid areas.

Water is essential for all the major chemical weathering reactions. To be effective in soil formation, water must penetrate the regolith. The seasonal rainfall distribution, evaporative losses, site topography, and soil permeability interact to determine how effectively precipitation can influence soil formation. The greater the depth of water penetration, the greater the depth of weathering of the soil and its development. Surplus water percolating through the soil profile transports soluble and suspended materials from the upper to the lower layers. It may also carry away soluble materials in the surface drainage waters. Thus, percolating water stimulates weathering reactions and helps differentiate soil horizons. Likewise, a deficiency of water is a major factor in determining the characteristics of soils of dry regions. Soluble salts are not leached from these soils, and in some cases they build up to levels that curtail plant growth. Soil profiles in arid and semi-arid regions are also apt to accumulate carbonates and certain types of expansive clays.

The direct influences of climate include:

- A shallow accumulation of lime in low rainfall areas as caliche
- Formation of acid soils in humid areas
- Erosion of soils on steep hillsides
- Deposition of eroded materials downstream
- Very intense chemical weathering, leaching, and erosion in warm and humid regions where soil does not freeze

Climate directly affects the rate of weathering and leaching. Wind moves sand and smaller particles, especially in arid regions where there is little plant cover. The type and amount of precipitation influence soil formation by affecting the movement of ions and particles through the soil, and aid in the development of different soil profiles. Soil profiles are more distinct in wet and cool climates, where organic materials may accumulate, than in wet and warm climates, where organic materials are rapidly consumed. The effectiveness of water in weathering parent rock material depends on seasonal and daily temperature fluctuations. Cycles of freezing and thawing constitute an effective mechanism which breaks up rocks and other consolidated materials.

Climate also indirectly influences soil formation through the effects of vegetation cover and biological activity, which modify the rates of chemical reactions in the soil.

Topography

The topography, or relief, is characterized by the inclination (slope), elevation, and orientation of the terrain. Topography determines the rate of precipitation or runoff and rate of formation or erosion of the surface soil profile. The topographical setting may either hasten or retard the work of climatic forces.

Steep slopes encourage rapid soil loss by erosion and allow less rainfall to enter the soil before running off and hence, little mineral deposition in lower profiles. In semiarid regions, the lower effective rainfall on steeper slopes also results in less complete vegetative cover, so there is less plant contribution to soil formation. For all of these reasons, steep slopes prevent the formation of soil from getting very far ahead of soil destruction. Therefore, soils on steep terrain tend to have rather shallow, poorly developed profiles in comparison to soils on nearby, more level sites.

In swales and depressions where runoff water tends to concentrate, the regolith is usually more deeply weathered and soil profile development is more advanced. However, in the lowest landscape positions, water may saturate the regolith to such a degree that drainage and aeration are restricted. Here, the weathering of some minerals and the decomposition of organic matter are retarded, while the loss of iron and manganese is accelerated. In such low-lying topography, special profile features characteristic of wetland soils may develop. Depressions allow the accumulation of water, minerals and organic matter and in the extreme, the resulting soils will be saline marshes or peat bogs. Intermediate topography affords the best conditions for the formation of an agriculturally productive soil.

Organisms

Soil is the most abundant ecosystem on Earth, but the vast majority of organisms in soil are microbes, a great many of which have not been described. There may be a population limit of around one billion cells per gram of soil, but estimates of the number of species vary widely from 50,000 per gram to over a million per gram of soil. The total number of organisms and species can vary widely according to soil type, location, and depth.

Plants, animals, fungi, bacteria and humans affect soil formation. Animals, soil mesofauna and micro-organisms mix soils as they form burrows and pores, allowing moisture and gases to move about. In the same way, plant roots open channels in soils. Plants with deep taproots can penetrate many metres through the different soil layers to bring up nutrients from deeper in the profile. Plants with fibrous roots that spread out near the soil surface have roots that are easily decomposed, adding organic matter. Micro-organisms, including fungi and bacteria, effect chemical exchanges between roots and soil and act as a reserve of nutrients.

Humans impact soil formation by removing vegetation cover with erosion as the result. Their tillage also mixes the different soil layers, restarting the soil formation process as less weathered material is mixed with the more developed upper layers.

Earthworms, ants and termites mix the soil as they burrow, significantly affecting soil formation. Earthworms ingest soil particles and organic residues, enhancing the availability of plant nutrients in the material that passes through their bodies. They aerate and stir the soil and increase the stability of soil aggregates, thereby assuring ready infiltration of water. In addition, as ants and termites build mounds, they transport soil materials from one horizon to another.

In general, the mixing of the soil by the activities of animals, sometimes called pedoturbation, tends to undo or counteract the tendency of other soil-forming processes that create distinct horizons. Termites and ants may also retard soil profile development by denuding large areas of soil around their nests, leading to increased loss of soil by erosion. Large animals such as gophers, moles, and prairie dogs bore into the lower soil horizons, bringing materials to the surface. Their tunnels are often open to the surface, encouraging the movement of water and air into the subsurface layers. In localized areas, they enhance mixing of the lower and upper horizons by creating, and later refilling, underground tunnels. Old animal burrows in the lower horizons often become filled with soil material from the overlying A horizon, creating profile features known as crotovinas.

Vegetation impacts soils in numerous ways. It can prevent erosion caused by excessive rain that might result from surface runoff. Plants shade soils, keeping them cooler and slow evaporation of soil moisture, or conversely, by way of transpiration, plants can cause soils to lose moisture. Plants can form new chemicals that can break down minerals and improve the soil structure. The type and amount of vegetation depends on climate, topography, soil characteristics, and biological factors. Soil factors such as density, depth, chemistry, pH, temperature and moisture greatly affect the type of plants that can grow in a given location. Dead plants and fallen leaves and stems begin their decomposition on the surface. There, organisms feed on them and mix the organic material with the upper soil layers; these added organic compounds become part of the soil formation process.

Human activities widely influence soil formation. For example, it is believed that Native Americans regularly set fires to maintain several large areas of prairie grasslands in Indiana and Michigan. In more recent times, human destruction of natural vegetation and subsequent tillage of the soil for crop production has abruptly modified soil formation. Likewise, irrigating an arid region of soil drastically influences the soil-forming factors, as does adding fertilizer and lime to soils of low fertility.

Time

Time is a factor in the interactions of all the above. While a mixture of sand, silt and clay constitute the texture of a soil and the aggregation of those components produces peds, the development of a distinct B horizon marks the development of a soil. With time, soils will evolve features that depend on the interplay of the prior listed soil-forming factors. It takes decades to several thousand years for a soil to develop a profile. That time period depends strongly on climate, parent material, relief, and biotic activity. For example, recently deposited material from a flood exhibits no soil development as there has not been enough time for the material to form a structure that further defines soil. The original soil surface is buried, and the formation process must begin anew for this deposit. Over time the soil will develop a profile that depends on the intensities of biota and climate. While a soil can achieve relative stability of its properties for extended periods, the soil life cycle ultimately ends in soil conditions that leave it vulnerable to erosion. Despite the inevitability of soil retrogression and degradation, most soil cycles are long.

Soil-forming factors continue to affect soils during their existence, even on "stable" landscapes that are long-enduring, some for millions of years. Materials are deposited on top or are blown or washed from the surface. With additions, removals and alterations, soils are always subject to new conditions. Whether these are slow or rapid changes depends on climate, topography and biological activity.

Physical Properties of Soils

The physical properties of soils, in order of decreasing importance, are texture, structure, density, porosity, consistency, temperature, colour and resistivity. Soil texture is determined by the relative proportion of the three kinds of soil mineral particles, called soil separates: sand, silt, and clay. At the next larger scale, soil structures called peds are created from the soil separates when iron oxides, carbonates, clay, silica and humus, coat particles and cause them to adhere into larger, relatively stable secondary structures. Soil density, particularly bulk density, is a measure of soil compaction. Soil porosity consists of the void part of the soil volume and is occupied by gases or water. Soil consistency is the ability of soil to stick together. Soil temperature and colour are self-defining. Resistivity refers to the resistance to conduction of electric currents and affects the rate of corrosion of metal and concrete structures. These properties may vary through the depth of a soil profile. Most of these properties determine the aeration of the soil and the ability of water to infiltrate and to be held within the soil.

Influence of Soil Texture Separates on Some Properties of Soils

Property/behavior	Sand	Silt	Clay
Water-holding capacity	Low	Medium to high	High
Aeration	Good	Medium	Poor
Drainage rate	High	Slow to medium	Very slow

Soil organic matter level	Low	Medium to high	High to medium
Decomposition of organic matter	Rapid	Medium	Slow
Warm-up in spring	Rapid	Moderate	Slow
Compactability	Low	Medium	High
Susceptibility to wind erosion	Moderate (High if fine sand)	High	Low
Susceptibility to water erosion	Low (unless fine sand)	High	Low if aggregated, otherwise high
Shrink/Swell Potential	Very Low	Low	Moderate to very high
Sealing of ponds, dams, and landfills	Poor	Poor	Good
Suitability for tillage after rain	Good	Medium	Poor
Pollutant leaching potential	High	Medium	Low (unless cracked)
Ability to store plant nutrients	Poor	Medium to High	High
Resistance to pH change	Low	Medium	High

Texture

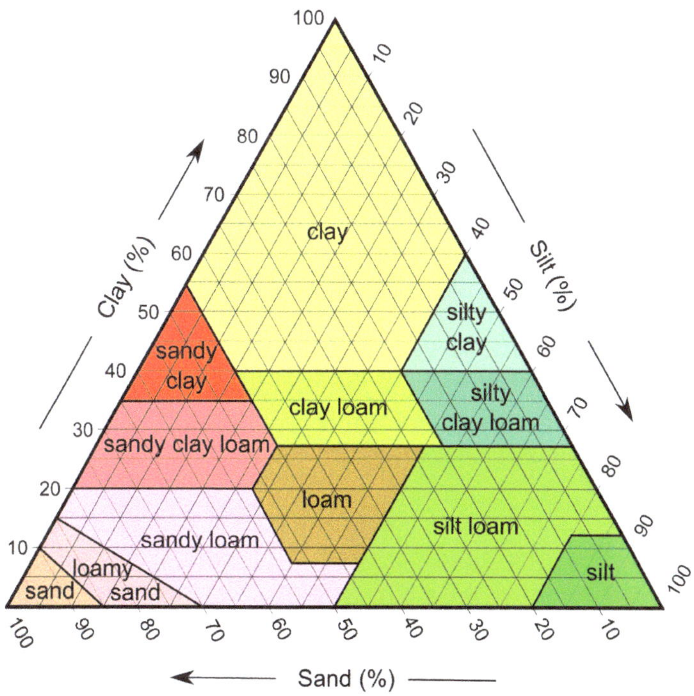

Soil types by clay, silt and sand composition as used by the USDA

Iron-rich soil near Paint Pots in Kootenay National Park, Canada

The mineral components of soil are sand, silt and clay, and their relative proportions determine a soil's texture. Properties that are influenced by soil texture, include porosity, permeability, infiltration, shrink-swell rate, water-holding capacity, and susceptibility to erosion. In the illustrated USDA textural classification triangle, the only soil in which neither sand, silt nor clay predominates is called "loam". While even pure sand, silt or clay may be considered a soil, from the perspective of food production a loam soil with a small amount of organic material is considered ideal. The mineral constituents of a loam soil might be 40% sand, 40% silt and the balance 20% clay by weight. Soil texture affects soil behaviour, in particular its retention capacity for nutrients and water.

Sand and silt are the products of physical and chemical weathering of the parent rock; clay, on the other hand, is a most often the product of the precipitation of the dissolved parent rock as a secondary mineral. It is the surface area to volume ratio (specific surface area) of soil particles and the unbalanced ionic charges within those that determine their role in the fertility of soil, as measured by its cation exchange capacity. Sand is least active, followed by silt; clay is the most active. Sand's greatest benefit to soil is that it resists compaction and increases a soil's porosity. Silt is mineralogically like sand but with its higher specific surface area it is more chemically active than sand. But it is the clay content of soil, with its very high specific surface area and generally large number of negative charges, that gives a soil its high retention capacity for water and nutrients. Clay soils also resist wind and water erosion better than silty and sandy soils, as the particles bond tightly to each other.

Sand is the most stable of the mineral components of soil; it consists of rock fragments, primarily quartz particles, ranging in size from 2.0 to 0.05 mm (0.0787 to 0.0020 in) in diameter. Silt ranges in size from 0.05 to 0.002 mm (0.002 to 0.00008 in). Clay cannot be resolved by optical microscopes as its particles are 0.002 mm (7.9×10^{-5} in) or less in diameter and a thickness of only 10 angstroms (10^{-10} m). In medium-textured soils, clay is often washed downward through the soil profile and accumulates in the subsoil.

Soil components larger than 2.0 mm (0.079 in) are classed as rock and gravel and are removed before determining the percentages of the remaining components and the texture class of the soil, but are included in the name. For example, a sandy loam soil with 20% gravel would be called gravelly sandy loam.

When the organic component of a soil is substantial, the soil is called organic soil rather than mineral soil. A soil is called organic if:

1. Mineral fraction is 0% clay and organic matter is 20% or more
2. Mineral fraction is 0% to 50% clay and organic matter is between 20% and 30%
3. Mineral fraction is 50% or more clay and organic matter 30% or more.

Structure

The clumping of the soil textural components of sand, silt and clay causes aggregates to form and the further association of those aggregates into larger units creates soil structures called peds (a contraction of the word pedolith). The adhesion of the soil textural components by organic substances, iron oxides, carbonates, clays, and silica, and the breakage of those aggregates from expansion-contraction, caused by freezing-thawing and wetting-drying cycles, shape soil into distinct geometric forms. The peds evolve into units which have various shapes, sizes and degrees of development. A soil clod, however, is not a ped but rather a mass of soil that results from mechanical disturbance of the soil. Soil structure affects aeration, water movement, conduction of heat, plant root growth and resistance to erosion. Water, in turn, has its strongest effect on soil structure due to its solution and precipitation of minerals and its effect on plant growth.

Soil structure often gives clues to its texture, organic matter content, biological activity, past soil evolution, human use, and the chemical and mineralogical conditions under which the soil formed. While texture is defined by the mineral component of a soil and is an innate property of the soil that does not change with agricultural activities, soil structure can be improved or destroyed by the choice and timing of farming practices.

Soil structural classes:

1. Types: Shape and arrangement of peds
    1. Platy: Peds are flattened one atop the other 1–10 mm thick. Found in the A-horizon of forest soils and lake sedimentation.
    2. Prismatic and Columnar: Prismlike peds are long in the vertical dimension, 10–100 mm wide. Prismatic peds have flat tops, columnar peds have rounded tops. Tend to form in the B-horizon in high sodium soil where clay has accumulated.

    3. Angular and subangular: Blocky peds are imperfect cubes, 5–50 mm, angular have sharp edges, subangular have rounded edges. Tend to form in the B-horizon where clay has accumulated and indicate poor water penetration.

    4. Granular and Crumb: Spheroid peds of polyhedrons, 1–10 mm, often found in the A-horizon in the presence of organic material. Crumb peds are more porous and are considered ideal.

2. Classes: Size of peds whose ranges depend upon the above type

    1. Very fine or very thin: <1 mm platy and spherical; <5 mm blocky; <10 mm prismlike.

    2. Fine or thin: 1–2 mm platy, and spherical; 5–10 mm blocky; 10–20 mm prismlike.

    3. Medium: 2–5 mm platy, granular; 10–20 mm blocky; 20-50 prismlike.

    4. Coarse or thick: 5–10 mm platy, granular; 20–50 mm blocky; 50–100 mm prismlike.

    5. Very coarse or very thick: >10 mm platy, granular; >50 mm blocky; >100 mm prismlike.

3. Grades: Is a measure of the degree of development or cementation within the peds that results in their strength and stability.

    1. Weak: Weak cementation allows peds to fall apart into the three textural constituents, sand, silt and clay.

    2. Moderate: Peds are not distinct in undisturbed soil but when removed they break into aggregates, some broken aggregates and little unaggregated material. This is considered ideal.

    3. Strong:Peds are distinct before removed from the profile and do not break apart easily.

    4. Structureless: Soil is entirely cemented together in one great mass such as slabs of clay or no cementation at all such as with sand.

At the largest scale, the forces that shape a soil's structure result from swelling and shrinkage that initially tend to act horizontally, causing vertically oriented prismatic peds. Clayey soil, due to its differential drying rate with respect to the surface, will induce horizontal cracks, reducing columns to blocky peds. Roots, rodents, worms, and freezing-thawing cycles further break the peds into a spherical shape.

At a smaller scale, plant roots extend into voids and remove water causing the open spaces to increase, thereby decreasing aggregate size. At the same time, roots, fungal hyphae, and earthworms create microscopic tunnels that break up peds.

At an even smaller scale, soil aggregation continues as bacteria and fungi exude sticky polysaccharides which bind soil into smaller peds. The addition of the raw organic matter that bacteria and fungi feed upon encourages the formation of this desirable soil structure.

At the lowest scale, the soil chemistry affects the aggregation or dispersal of soil particles. The clay particles contain polyvalent cations which give the faces of clay layers localized negative charges. At the same time, the edges of the clay plates have a slight positive charge, thereby allowing the edges to adhere to the negative charges on the faces of other clay particles or to flocculate (form clumps). On the other hand, when monovalent ions, such as sodium, invade and displace the polyvalent cations, they weaken the positive charges on the edges, while the negative surface charges are relatively strengthened. This leaves negative charge on the clay faces that repel other clay, causing the particles to push apart, and by doing,the flocculation of clay particles into larger, open assemblages. As a result, the clay disperses and settles into voids between peds, causing those to close. In this way the open structure of the soil is destroyed and the soil is made impenetrable to air and water. Such sodic soil tends to form columnar structures near the surface.

Density

Soil particle density is typically 2.60 to 2.75 grams per cm^3 and is usually unchanging for a given soil. Soil particle density is lower for soils with high organic matter content, and is higher for soils with high iron-oxides content. Soil bulk density is equal to the dry mass of the soil divided by the volume of the soil; i.e., it includes air space and organic materials of the soil volume. The soil bulk density of cultivated loam is about 1.1 to 1.4 g/cm^3 (for comparison water is 1.0 g/cm^3). Soil bulk density is highly variable for a given soil. A lower bulk density by itself does not indicate suitability for plant growth due to the influence of soil texture and structure. A high bulk density is indicative of either soil compaction or high sand content. Soil bulk density is inherently always less than the soil particle density.

Representative bulk densities of soils. The percentage pore space was calculated using 2.7 g/cm^3 for particle density except for the peat soil, which is estimated.

Soil treatment and identification	Bulk density g/cm^3	Pore space %
Tilled surface soil of a cotton field	1.3	51
Trafficked inter-rows where wheels passed surface	1.67	37
Traffic pan at 25 cm deep	1.7	36
Undisturbed soil below traffic pan, clay loam	1.5	43

Rocky silt loam soil under aspen forest	1.62	40
Loamy sand surface soil	1.5	43
Decomposed peat	0.55	65

Porosity

Pore space is that part of the bulk volume of soil that is not occupied by either mineral or organic matter but is open space occupied by either gases or water. In a productive, medium-textured soil the total pore space is typically about 50% of the soil volume. Pore size varies considerably; the smallest pores (cryptopores; <0.1 μm) hold water too tightly for use by plant roots; plant-available water is held in ultramicropores, micropores and mesopores (0.1-75 μm); and macropores (>75 μm) are generally air-filled when the soil is at field capacity.

Soil texture determines total volume of the smallest pores; clay soils have smaller pores, but more total pore space than sands. Soil structure has a strong influence on the larger pores that affect soil aeration, water infiltration and drainage. Tillage has the short-term benefit of temporarily increasing the number of pores of largest size, but these can be rapidly degraded by the destruction of soil aggregation.

The pore size distribution affects the ability of plants and other organisms to access water and oxygen; large, continuous pores allow rapid transmission of air, water and dissolved nutrients through soil, and small pores store water between rainfall or irrigation events. Pore size variation also compartmentalizes the soil pore space such that many microorganisms are not in direct competition with one another, which may explain not only the large number of species present, but the fact that functionally redundant microorganisms (organisms with the same ecological niche) can co-exist within the same soil.

Consistency

Consistency is the ability of soil to stick to itself or to other objects (cohesion and adhesion respectively) and its ability to resist deformation and rupture. It is of approximate use in predicting cultivation problems and the engineering of foundations. Consistency is measured at three moisture conditions: air-dry, moist, and wet. In those conditions the consistency quality depends upon the clay content. In the wet state, the two qualities of stickiness and plasticity are assessed. A soil's resistance to fragmentation and crumbling is assessed in the dry state by rubbing the sample. Its resistance to shearing forces is assessed in the moist state by thumb and finger pressure. Additionally, the cemented consistency depends on cementation by substances other than clay, such as calcium carbonate, silica, oxides and salts; moisture content has little effect on its assessment. The measures of consistency border on subjective compared to other measures such as pH, since they employ the apparent feel of the soil in those states.

The terms used to describe the soil consistency in three moisture states and a last not affected by the amount of moisture are as follows:

1. Consistency of Dry Soil: loose, soft, slightly hard, hard, very hard, extremely hard
2. Consistency of Moist Soil: loose, very friable, friable, firm, very firm, extremely firm
3. Consistency of Wet Soil: nonsticky, slightly sticky, sticky, very sticky; nonplastic, slightly plastic, plastic, very plastic
4. Consistency of Cemented Soil: weakly cemented, strongly cemented, indurated (requires hammer blows to break up)

Soil consistency is useful in estimating the ability of soil to support buildings and roads. More precise measures of soil strength are often made prior to construction.

Temperature

Soil temperature depends on the ratio of the energy absorbed to that lost. Soil has a temperature range between -20 to 60 °C. Soil temperature regulates seed germination, plant and root growth and the availability of nutrients. Below 50 cm (20 in), soil temperature seldom changes and can be approximated by adding 1.8 °C (2 °F) to the mean annual air temperature. Soil temperature has important seasonal, monthly and daily variations. Fluctuations in soil temperature are much lower with increasing soil depth. Heavy mulching (a type of soil cover) can slow the warming of soil, and, at the same time, reduce fluctuations in surface temperature.

Most often, agricultural activities must adapt to soil temperatures by:

1. maximizing germination and growth by timing of planting
2. optimizing use of anhydrous ammonia by applying to soil below 10 °C (50 °F)
3. preventing heaving and thawing due to frosts from damaging shallow-rooted crops
4. preventing damage to desirable soil structure by freezing of saturated soils
5. improving uptake of phosphorus by plants

Soil temperatures can be raised by drying soils or the use of clear plastic mulches. Organic mulches slow the warming of the soil.

There are various factors that affect soil temperature, such as water content, soil color, and relief (slope, orientation, and elevation), and soil cover (shading and insulation). The color of the ground cover and its insulating properties have a strong influence on soil temperature. Whiter soil tends to have a higher albedo than blacker soil cover, which encourages whiter soils to have lower soil temperatures. The specific heat of soil is the energy required to raise the temperature of soil by 1 °C. The specific heat of soil increases as water content increases, since the heat capacity of water is greater than that of dry soil. The specific heat of pure water is ~ 1 calorie per gram, the specific heat of dry soil is ~ 0.2 calories per gram, hence, the specific heat of wet soil is ~ 0.2 to 1 calories per gram. Also, a tremendous energy (~540 cal/g) is required to evaporate water (known as the heat of vaporization). As such, wet soil usually warms more slowly than dry soil - wet surface soil is typically 3 to 6 °C colder than dry surface soil.

Soil heat flux refers to the rate at which heat energy moves through the soil in response to a temperature difference between two points in the soil. The heat flux density is the amount of energy that flows through soil per unit area per unit time and has both magnitude and direction. For the simple case of conduction into or out of the soil in the vertical direction, which is most often applicable the heat flux density is:

$$q_x = -k\frac{\delta T}{\delta x}$$

In SI units

q is the heat flux density, in SI the units are W·m^{-2}

is the soils' conductivity, W·m^{-1}·K^{-1}. The thermal conductivity is sometimes a constant, otherwise an average value of conductivity for the soil condition between the surface and the point at depth is used.

δT is the temperature difference (temperature gradient) between the two points in the soil between which the heat flux density is to be calculated. In SI the units are kelvin, K.

δx is the distance between the two points within the soil, at which the temperatures are measured and between which the heat flux density is being calculated. In SI the units are meters m, and where x is measured positive downward.

Heat flux is in the direction opposite the temperature gradient, hence the minus sign. That is to say, if the temperature of the surface is higher than at depth x the negative sign will result in a positive value for the heat flux q, and which is interpreted as the heat being conducted into the soil.

Component	**Thermal Conductivity (W·m-1·K-1)**
Quartz	8.8
Clay	2.9

Organic matter	0.25
Water	0.57
Ice	2.4
Air	0.025
Dry soil	0.2-0.4
Wet soil	1-3

Soil temperature is important for the survival and early growth of seedlings. Soil temperatures affect the anatomical and morphological character of root systems. All physical, chemical, and biological processes in soil and roots are affected in particular because of the increased viscosities of water and protoplasm at low temperatures. In general, climates that do not preclude survival and growth of white spruce above ground are sufficiently benign to provide soil temperatures able to maintain white spruce root systems. In some northwestern parts of the range, white spruce occurs on permafrost sites and although young unlignified roots of conifers may have little resistance to freezing, less than half of the "secondary mature" root system of white spruce was killed by exposure to a temperature of 23.3 °C in multiple year experiment with containerized trees from local nurseries in Massachusetts.

Optimum temperatures for tree root growth range between 10 °C and 25 °C in general and for spruce in particular. In 2-week-old white spruce seedlings that were then grown for 6 weeks in soil at temperatures of 15 °C, 19 °C, 23 °C, 27 °C, and 31 °C; shoot height, shoot dry weight, stem diameter, root penetration, root volume, and root dry weight all reached maxima at 19 °C.

However, whereas strong positive relationships between soil temperature (5 °C to 25 °C) and growth have been found in trembling aspen and balsam poplar, white and other spruce species have shown little or no changes in growth with increasing soil temperature. Such insensitivity to soil low temperature may be common among a number of western and boreal conifers.

Color

Soil colour is often the first impression one has when viewing soil. Striking colours and contrasting patterns are especially noticeable. The Red River of the South carries sediment eroded from extensive reddish soils like Port Silt Loam in Oklahoma. The Yellow River in China carries yellow sediment from eroding loess soils. Mollisols in the Great Plains of North America are darkened and enriched by organic matter. Podsols in boreal forests have highly contrasting layers due to acidity and leaching.

In general, color is determined by the organic matter content, drainage conditions, and degree of oxidation. Soil color, while easily discerned, has little use in predicting soil characteristics. It is of use in distinguishing boundaries within a soil profile, determining the origin of a soil's parent material, as an indication of wetness and waterlogged conditions, and as a qualitative means of measuring organic, salt and carbonate contents of soils. Color is recorded in the Munsell color system as for instance 10YR3/4 *Dusky Red.*

Soil color is primarily influenced by soil mineralogy. Many soil colours are due to various iron minerals. The development and distribution of colour in a soil profile result from chemical and biological weathering, especially redox reactions. As the primary minerals in soil parent material weather, the elements combine into new and colourful compounds. Iron forms secondary minerals of a yellow or red colour, organic matter decomposes into black and brown compounds, and manganese, sulfur and nitrogen can form black mineral deposits. These pigments can produce various colour patterns within a soil. Aerobic conditions produce uniform or gradual colour changes, while reducing environments (anaerobic) result in rapid colour flow with complex, mottled patterns and points of colour concentration.

Resistivity

Soil resistivity is a measure of a soil's ability to retard the conduction of an electric current. The electrical resistivity of soil can affect the rate of galvanic corrosion of metallic structures in contact with the soil. Higher moisture content or increased electrolyte concentration can lower resistivity and increase conductivity, thereby increasing the rate of corrosion. Soil resistivity values typically range from about 2 to 1000 Ω·m, but more extreme values are not unusual.

Soil Water

Water that enters a field is removed from a field by runoff, drainage, evaporation or transpiration. Runoff is the water that flows on the surface to the edge of the field; drainage is the water that flows through the soil downward or toward the edge of the field underground; evaporative water loss from a field is that part of the water that evaporates into the atmosphere directly from the field's surface; transpiration is the loss of water from the field by its evaporation from the plant itself.

Water affects soil formation, structure, stability and erosion but is of primary concern with respect to plant growth. Water is essential to plants for four reasons:

1. It constitutes 80%-95% of the plant's protoplasm.
2. It is essential for photosynthesis.
3. It is the solvent in which nutrients are carried to, into and throughout the plant.
4. It provides the turgidity by which the plant keeps itself in proper position.

In addition, water alters the soil profile by dissolving and re-depositing minerals, often at lower levels, and possibly leaving the soil sterile in the case of extreme rainfall and drainage. In a loam soil, solids constitute half the volume, gas one-quarter of the volume, and water one-quarter of the volume of which only half of which will be available to most plants.

A flooded field will drain the gravitational water under the influence of gravity until water's adhesive and cohesive forces resist further drainage at which point it is said to have reached field capacity. At that point, plants must apply suction to draw water from a soil. The water that plants may draw from the soil is called the available water. Once the available water is used up the remaining moisture is called unavailable water as the plant cannot produce sufficient suction to draw that water in. A plant must produce suction that increases from zero for a flooded field to 1/3 bar at field dry condition (one bar is a little less than one atmosphere pressure). At 15 bar suction, wilting percent, seeds will not germinate, plants begin to wilt and then die. Water moves in soil under the influence of gravity, osmosis and capillarity. When water enters the soil, it displaces air from some of the pores, since air content of a soil is inversely related to its water content.

The rate at which a soil can absorb water depends on the soil and its other conditions. As a plant grows, its roots remove water from the largest pores first. Soon the larger pores hold only air, and the remaining water is found only in the intermediate- and smallest-sized pores. The water in the smallest pores is so strongly held to particle surfaces that plant roots cannot pull it away. Consequently, not all soil water is available to plants. When saturated, the soil may lose nutrients as the water drains. Water moves in a draining field under the influence of pressure where the soil is locally saturated and by capillarity pull to dryer parts of the soil. Most plant water needs are supplied from the suction caused by evaporation from plant leaves and 10% is supplied by "suction" created by osmotic pressure differences between the plant interior and the soil water. Plant roots must seek out water. Insufficient water will damage the yield of a crop. Most of the available water is used in transpiration to pull nutrients into the plant.

Water Retention Forces

Water is retained in a soil when the adhesive force of attraction that water's hydrogen atoms have for the oxygen of soil particles is stronger than the cohesive forces that water's hydrogen feels for other water oxygen atoms. When a field is flooded, the soil pore space is completely filled by water. The field will drain under the force of gravity until it reaches what is called field capacity, at which point the smallest pores are filled with water and the largest with water and gases. The total amount of water held when field capacity is reached is a function of the specific surface area of the soil particles. As a result, high clay and high organic soils have higher field capacities. The total force required to pull or push water out of soil is termed suction and usually expressed in units of bars (10^5 pascal) which is just a little less than one-atmosphere pressure. Alternatively, the terms "tension" or "moisture potential" may be used.

Moisture Classification

The forces with which water is held in soils determine its availability to plants. Forces of adhesion hold water strongly to mineral and humus surfaces and less strongly to itself by cohesive forces. A plant's root may penetrate a very small volume of water that is adhering to soil and be initially able to draw in water that is only lightly held by the cohesive forces. But as the droplet is drawn down, the forces of adhesion of the water for the soil particles make reducing the volume of water increasingly difficult until the plant cannot produce sufficient suction to use the remaining water. The remaining water is considered unavailable. The amount of available water depends upon the soil texture and humus amounts and the type of plant attempting to use the water. Cacti, for example, can produce greater suction than can agricultural crop plants.

The following description applies to a loam soil and agricultural crops. When a field is flooded, it is said to be saturated and all available air space is occupied by water. The suction required to draw water into a plant root is zero. As the field drains under the influence of gravity (drained water is called gravitational water or drain-able water), the suction a plant must produce to use such water increases to 1/3 bar. At that point, the soil is said to have reached field capacity, and plants that use the water must produce increasingly higher suction, finally up to 15 bar. At 15 bar suction, the soil water amount is called wilting percent. At that suction the plant cannot sustain its water needs as water is still being lost from the plant by transpiration; the plant's turgidity is lost, and it wilts. The next level, called air-dry, occurs at 1000 bar suction. Finally the oven dry condition is reached at 10,000 bar suction. All water below wilting percentage is called unavailable water.

Soil Moisture Content

When the soil moisture content is optimal for plant growth, the water in the large and intermediate size pores can move about in the soil and be easily used by plants. The amount of water remaining in a soil drained to field capacity and the amount that is available are functions of the soil type. Sandy soil will retain very little water, while clay will hold the maximum amount. The time required to drain a field from flooded condition for a clay loam that begins at 43% water by weight to a field capacity of 21.5% is six days, whereas a sand loam that is flooded to its maximum of 22% water will take two days to reach field capacity of 11.3% water. The available water for the clay loam might be 11.3% whereas for the sand loam it might be only 7.9% by weight.

Wilting point, field capacity, and available water capacity of various soil textures						
Soil Texture	Wilting Point		Field Capacity		Available water capacity	
	Water per foot of soil depth		Water per foot of soil depth		Water per foot of soil depth	
	%	in.	%	in.	%	in.
Medium sand	1.7	0.3	6.8	1.2	5.1	0.9

Fine sand	2.3	0.4	8.5	1.5	6.2	1.1
Sandy loam	3.4	0.6	11.3	2.0	7.9	1.4
Fine sandy loam	4.5	0.8	14.7	2.6	10.2	1.8
Loam	6.8	1.2	18.1	3.2	11.3	2.0
Silt loam	7.9	1.4	19.8	3.5	11.9	2.1
Clay loam	10.2	1.8	21.5	3.8	11.3	2.0
Clay	14.7	2.6	22.6	4.0	7.9	1.4

The above are average values for the soil textures as the percentages of sand, silt and clay vary.

Water Flow in Soils

Water moves through soil due to the force of gravity, osmosis and capillarity. At zero to one-third bar suction, water is pushed through soil from the point of its application under the force of gravity and the pressure gradient created by the pressure of the water; this is called saturated flow. At higher suction, water movement is pulled by capillarity from wetter toward dryer soil. This is caused by water's adhesion to soil solids, and is called unsaturated flow.

Water infiltration and movement in soil is controlled by six factors:

1. Soil texture
2. Soil structure. Fine-textured soils with granular structure are most favourable to infiltration of water.
3. The amount of organic matter. Coarse matter is best and if on the surface helps prevent the destruction of soil structure and the creation of crusts.
4. Depth of soil to impervious layers such as hardpans or bedrock
5. The amount of water already in the soil
6. Soil temperature. Warm soils take in water faster while frozen soils may not be able to absorb depending on the type of freezing.

Water infiltration rates range from 0.25 cm (0.098 in) per hour for high clay soils to 2.5 cm (0.98 in) per hour for sand and well stabilised and aggregated soil structures. Water flows through the ground unevenly, called "gravity fingers", because of the surface tension between water particles. Tree roots create paths for rainwater flow through soil by breaking though soil including clay layers: one study showed roots increasing infiltration of water by 153% and another study showed an increase by 27 times. Flooding temporarily increases soil permeability in river beds, helping to recharge aquifers.

Saturated Flow

Water applied to a soil is pushed by pressure gradients from the point of its application where it is saturated locally, to less saturated areas. Once soil is completely wetted, any more water will move downward, or percolate, carrying with it clay, humus and nutrients, primarily cations, out of the range of plant roots. In order of decreasing solubility, the leached nutrients are:

- Calcium
- Magnesium, Sulfur, Potassium; depending upon soil composition
- Nitrogen; usually little, unless nitrate fertiliser was applied recently
- Phosphorus; very little as its forms in soil are of low solubility.

In the United States percolation water due to rainfall ranges from zero inches just east of the Rocky Mountains to twenty or more inches in the Appalachian Mountains and the north coast of the Gulf of Mexico.

Unsaturated Flow

At suctions less than one-third bar, water moves in all directions via unsaturated flow at a rate that is dependent on the square of the diameter of the water-filled pores. Water is pulled by capillary action due to the adhesion force of water to the soil solids, producing a suction gradient from wet towards drier soil. Doubling the diameter of the pores increases the flow rate by a factor of four. Large pores drained by gravity and not filled with water do not greatly increase the flow rate for unsaturated flow. Water flow is primarily from coarse-textured soil into fine-textured soil and is slowest in fine-textured soils such as clay.

Water Uptake by Plants

Of equal importance to the storage and movement of water in soil is the means by which plants acquire it and their nutrients. Ninety percent of water is taken up by plants as passive absorption caused by the pulling force of water evaporating (transpiring) from the long column of water that leads from the plant's roots to its leaves. In addition, the high concentration of salts within plant roots creates an osmotic pressure gradient that pushes soil water into the roots. Osmotic absorption becomes more important during times of low water transpiration caused by lower temperatures (for example at night) or high humidity. It is the process that causes guttation.

Root extension is vital for plant survival. A study of a single winter rye plant grown for four months in one cubic foot of loam soil showed that the plant developed 13,800,000 roots, a total of 385 miles in length with 2,550 square feet in surface area; and 14 billion hair roots of 6,600 miles total length and 4,320 square feet total area; for a total surface area of 6,870 square feet (83 ft squared). The total surface area of the loam soil was estimated to be 560,000 square feet. In other words, the roots were in contact with only 1.2% of the soil.

Roots must seek out water as the unsaturated flow of water in soil can move only at a rate of up to 2.5 cm (one inch) per day; as a result they are constantly dying and growing as they seek out high concentrations of soil moisture. Insufficient soil moisture, to the point of causing wilting, will cause permanent damage and crop yields will suffer. When grain sorghum was exposed to soil suction as low as 13.0 bar during the seed head emergence through bloom and seed set stages of growth, its production was reduced by 34%.

Consumptive use and Water Efficiency

Only a small fraction (0.1% to 1%) of the water used by a plant is held within the plant. The majority is ultimately lost via transpiration, while evaporation from the soil surface is also substantial. Transpiration plus evaporative soil moisture loss is called evapotranspiration. Evapotranspiration plus water held in the plant totals to consumptive use, which is nearly identical to evapotranspiration.

The total water used in an agricultural field includes runoff, drainage and consumptive use. The use of loose mulches will reduce evaporative losses for a period after a field is irrigated, but in the end the total evaporative loss will approach that of an uncovered soil. The benefit from mulch is to keep the moisture available during the seedling stage. Water use efficiency is measured by transpiration ratio, which is the ratio of the total water transpired by a plant to the dry weight of the harvested plant. Transpiration ratios for crops range from 300 to 700. For example, alfalfa may have a transpiration ratio of 500 and as a result 500 kilograms of water will produce one kilogram of dry alfalfa.

Soil Atmosphere

The atmosphere of soil is radically different from the atmosphere above. The consumption of oxygen, by microbes and plant roots and their release of carbon dioxide, decrease oxygen and increase carbon dioxide concentration. Atmospheric CO_2 concentration is 0.04%, but in the soil pore space it may range from 10 to 100 times that level. At extreme levels CO_2 is toxic. In addition, the soil voids are saturated with water vapour. Adequate porosity is necessary, not just to allow the penetration of water, but also to allow gases to diffuse in and out. Movement of gases is by diffusion from high concentrations to lower. Oxygen diffuses in and is consumed and excess levels of carbon dioxide, diffuse out with other gases as well as water. Soil texture and structure strongly affect soil porosity and gas diffusion. It is the total pore space (porosity) of soil not the pore size that determines the rate of diffusion of gases into and out of soil. A Platy soil structure and compacted soils (low porosity) impede gas flow, and a deficiency of oxygen may encourage anaerobic bacteria to reduce (strip oxygen) from nitrate NO_3 to the gases N_2, N_2O, and NO, which are then lost to the atmosphere, thereby depleting the soil of nitrogen. Aerated soil is also a net sink of methane CH_4 but a net producer of greenhouse gases when soils are depleted of oxygen and subject to elevated temperatures.

Composition of Soil Particles

Soil particles can be classified by their chemical composition (mineralogy) as well as their size. The particle size distribution of a soil, its texture, determines many of the properties of that soil, but the mineralogy of those particles can strongly modify those properties. The mineralogy of the finest soil particles, clay, is especially important.

Gravel, Sand and Silt

Gravel, sand and silt are the larger soil particles, and their mineralogy is often inherited from the parent material of the soil, but may include products of weathering (such as concretions of calcium carbonate or iron oxide), or residues of plant and animal life (such as silica phytoliths). Quartz is the most common mineral in the sand or silt fraction as it is resistant to chemical weathering; other common minerals are feldspars, micas and ferromagnesian minerals such as pyroxenes, amphiboles and olivines.

Mineral Colloids; Soil Clays

Due to its high specific surface area and its unbalanced negative charges, clay is the most active mineral component of soil. It is a colloidal and most often a crystalline material. In soils, clay is a soil textural class and is defined in a physical sense as any mineral particle less than 2 μm (8×10^{-5} in) in effective diameter. Many soil minerals, such as gypsum, carbonates, or quartz, are small enough to be classified as clay based on their physical size, but chemically they do not afford the same utility as do clay minerals. Chemically, clay is a range of minerals with certain reactive properties.

Clay was once thought to be very small particles of quartz, feldspar, mica, hornblende or augite, but it is now known to be (with the exception of mica-based clays) a precipitate with a mineralogical composition that is dependent on but different from its parent materials and is classed as a secondary mineral. The type of clay that is formed is a function of the parent material and the composition of the minerals in solution. Clay minerals continue to be formed as long as the soil exists. Mica-based clays result from a modification of the primary mica mineral in such a way that it behaves and is classed as a clay. Most clays are crystalline, but some are amorphous. The clays of a soil are a mixture of the various types of clay, but one type predominates.

There are four groups of clay: layer silicates; crystalline chain silicates; metal oxides and hydroxides and oxy-oxides; and amorphous; and allophanes. Most clays are crystalline and most are made up of three or four planes of oxygen held together by planes of aluminium and silicon by way of ionic bonds that together form a single layer of clay. The spatial arrangement of the oxygen atoms determines clay's structure. Half of the weight of clay is oxygen, but on a volume basis oxygen is ninety percent. The layers of clay are sometimes held together through hydrogen bonds or potassium bridges and as a result will swell less in the presence of water. Other clays, such as montmorillonite, have layers that are loosely attached and will swell greatly when water intervenes between the layers.

There are four groups of clays:

1. Layer Crystalline alumino-silica clays: montmorillonite, illite, vermiculite, chlorite, kaolinite.
2. Crystalline Chain carbonate and sulfate minerals: calcite ($CaCO_3$), dolomite ($CaMg(CO_3)_2$) and gypsum ($CaSO_4 \cdot 2H_2O$).
3. Amorphous clays: young mixtures of silica (SiO_2-OH) and alumina ($Al(OH)_3$) which have not had time to form regular crystals.
4. Sesquioxide clays: old, highly leached clays which result in oxides of iron, aluminium and titanium.

Alumino-silica Clays

Alumino-silica clays are characterised by their regular crystalline structure. Oxygen in ionic bonds with silicon forms a tetrahedral coordination (silicon at the center) which in turn forms sheets of silica. Two sheets of silica are bonded together by a plane of aluminium which forms an octahedral coordination, called alumina, with the oxygens of the silica sheet above and that below it. Hydroxyl ions (OH^-) sometimes substitute for oxygen. During the clay formation process, Al^{3+} may substitute for Si^{4+} in the silica layer, and as much as one fourth of the aluminium Al^{3+} may be substituted by Zn^{2+}, Mg^{2+} or Fe^{2+} in the alumina layer. The substitution of lower-valence cations for higher-valence cations (isomorphous substitution) gives clay a local negative charge on an oxygen atom that attracts and holds water and positively charged soil cations, some of which are of value for plant growth. Isomorphous substitution occurs during the clay's formation and does not change with time.

- Montmorillonite clay is made of four planes of oxygen with two silicon and one central aluminium plane intervening. The alumino-silicate montmorillonite clay is said to have a 2:1 ratio of silicon to aluminium. The seven planes together form a single crystal of montmorillonite. The crystals are weakly held together and water may intervene, causing the clay to swell up to ten times its dry volume. It occurs in soils which have had little leaching, hence it is found in arid regions. As the crystals are not bonded face to face, the entire surface is exposed and available for surface reactions, hence it has a high cation exchange capacity (CEC).

- Illite is a 2:1 clay similar in structure to montmorillonite but has potassium bridges between the faces of the clay crystals and the degree of swelling depends on the degree of weathering of the potassium. The active surface area is reduced due to the potassium bonds. Illite originates from the modification of mica, a primary mineral. It is often found together with montmorillonite and its primary minerals. It has moderate CEC.

- Vermiculite is a mica-based clay similar to illite, but the crystals of clay are held together more loosely by hydrated magnesium and it will swell, but not as much as does montmorillonite. It has very high CEC.

- Chlorite is similar to vermiculite, but the loose bonding by occasional hydrated magnesium, as in vermiculite, is replaced by a hydrated magnesium sheet, that firmly bonds the planes above and below it. It has two planes of silicon, one of aluminium and one of magnesium; hence it is a 2:2 clay. Chlorite does not swell and it has low CEC.

- Kaolinite is very common, highly weathered clay, and more common than montmorillonite in acid soils. It has one silica and one alumina plane per crystal; hence it is a 1:1 type clay. One plane of silica of montmorillonite is dissolved and is replaced with hydroxyls, which produces strong hydrogen bonds to the oxygen in the next crystal of clay. As a result, kaolinite does not swell in water and has a low specific surface area, and as almost no isomorphous substitution has occurred it has a low CEC. Where rainfall is high, acid soils selectively leach more silica than alumina from the original clays, leaving kaolinite. Even heavier weathering results in sesquioxide clays.

silica-sesquioxide

Crystalline Chain Clays

The carbonate and sulfate minerals are much more soluble and hence are found primarily in desert soils where leaching is less active.

Amorphous Clays

Amorphous clays are young, and commonly found in volcanic ash. They are mixtures of alumina and silica which have not formed the ordered crystal shape of alumino-silica clays which time would provide. The majority of their negative charges originates from hydroxyl ions, which can gain or lose a hydrogen ion (H^+) in response to soil pH, in such way was as to buffer the soil pH. They may have either a negative charge provided by the attached hydroxyl ion (OH^-), which can attract a cation, or lose the hydrogen of the hydroxyl to solution and display a positive charge which can attract anions. As a result, they may display either high CEC in an acid soil solution, or high anion exchange capacity in a basic soil solution.

Sesquioxide Clays

Sesquioxide clays are a product of heavy rainfall that has leached most of the silica from alumino-silica clay, leaving the less soluble oxides iron hematite (Fe_2O_3), iron hydroxide ($Fe(OH)_3$), aluminium hydroxide gibbsite ($Al(OH)_3$), hydrated manganese birnessite (MnO_2). It takes hundreds of thousands of years of leaching to create sesquioxide clays. *Sesqui* is Latin for "one and one-half": there are three parts oxygen to two parts iron or aluminium; hence the ratio is one and one-half (not true for all). They are hydrated and act as either amorphous or crystalline. They are not sticky and do not swell, and soils high in them behave much like sand and can rapidly pass water. They are able to hold large quantities of phosphates. Sesquioxides have low CEC but are able to hold anions as well as cations. Such soils range from yellow to red in colour. Such clays tend to hold phosphorus so tightly that it is unavailable for absorption by plants.

Organic Colloids

Humus is the final state of decomposition of organic matter. While it may linger for a thousand years, on the larger scale of the age of the mineral soil components, it is temporary. It is composed of the very stable lignins (30%) and complex sugars (polyuronides, 30%), proteins (30%), waxes, and fats that are resistant to breakdown by microbes. Its chemical assay is 60% carbon, 5% nitrogen, some oxygen and the remainder hydrogen, sulfur, and phosphorus. On a dry weight basis, the CEC of humus is many times greater than that of clay.

Carbon and Terra Preta

In the extreme environment of high temperatures and the leaching caused by the heavy rain of tropical rain forests, the clay and organic colloids are largely destroyed.

The heavy rains wash the alumino-silicate clays from the soil leaving only sesquioxide clays of low CEC. The high temperatures and humidity allow bacteria and fungi to virtually dissolve any organic matter on the rain-forest floor overnight and much of the nutrients are volatilized or leached from the soil and lost. However, carbon in the form of charcoal is far more stable than soil colloids and is capable of performing many of the functions of the soil colloids of sub-tropical soils. Soil containing substantial quantities of charcoal, of an anthropogenic origin, is called terra preta. Research into terra preta is still young but is promising. Fallow periods "on the Amazonian Dark Earths can be as short as 6 months, whereas fallow periods on oxisols are usually 8 to 10 years long".

Soil Chemistry

The chemistry of a soil determines its ability to supply available plant nutrients and affects its physical properties and the health of its microbial population. In addition, a soil's chemistry also determines its corrosivity, stability, and ability to absorb pollutants and to filter water. It is the surface chemistry of mineral and organic colloids that determines soil's chemical properties. "A colloid is a small, insoluble, nondiffusible particle larger than a molecule but small enough to remain suspended in a fluid medium without settling. Most soils contain organic colloidal particles called humus as well as the inorganic colloidal particles of clays." The very high specific surface area of colloids and their net charges, gives soil its ability to hold and release ions. Negatively charged sites on colloids attract and release cations in what is referred to as cation exchange. Cation-exchange capacity (CEC) is the amount of exchangeable cations per unit weight of dry soil and is expressed in terms of milliequivalents of positively charged ions per 100 grams of soil (or centimoles of positive charge per kilogram of soil; $cmol_c/kg$). Similarly, positively charged sites on colloids can attract and release anions in the soil giving the soil anion exchange capacity (AEC).

Cation and Anion Exchange

The cation exchange, that takes place between colloids and soil water, buffers (moderates) soil pH, alters soil structure, and purifies percolating water by adsorbing cations of all types, both useful and harmful.

The negative or positive charges on colloid particles make them able to hold cations or anions, respectively, to their surfaces. The charges result from four sources.

1. Isomorphous substitution occurs in clay during its formation, when lower-valence cations substitute for higher-valence cations in the crystal structure. Substitutions in the outermost layers are more effective than for the innermost layers, as the charge strength drops off as the square of the distance. The net result is oxygen atoms with net negative charge and the ability to attract cations.

2. Edge-of-clay oxygen atoms are not in balance ionically as the tetrahedral and octahedral structures are incomplete.

3. Hydroxyls may substitute for oxygens of the silica layers. When the hydrogens of the clay hydroxyls are ionised into solution, they leave the oxygen with a negative charge.

4. Hydrogens of humus hydroxyl groups may be ionised into solution, leaving an oxygen with a negative charge.

Cations held to the negatively charged colloids resist being washed downward by water and out of reach of plants' roots, thereby preserving the fertility of soils in areas of moderate rainfall and low temperatures.

There is a hierarchy in the process of cation exchange on colloids, as they differ in the strength of adsorption by the colloid and hence their ability to replace one another. If present in equal amounts in the soil water solution:

Al^{3+} replaces H^+ replaces Ca^{2+} replaces Mg^{2+} replaces K^+ same as NH^{4+} replaces Na^+

If one cation is added in large amounts, it may replace the others by the sheer force of its numbers. This is called mass action. This is largely what occurs with the addition of fertiliser.

As the soil solution becomes more acidic (low pH, and an abundance of H^+), the other cations more weakly bound to colloids are pushed into solution as hydrogen ions occupy those sites. A low pH may cause hydrogen of hydroxyl groups to be pulled into solution, leaving charged sites on the colloid available to be occupied by other cations. This ionisation of hydroxyl groups on the surface of soil colloids creates what is described as pH-dependent charges. Unlike permanent charges developed by isomorphous substitution, pH-dependent charges are variable and increase with increasing pH. Freed cations can be made available to plants but are also prone to be leached from the soil, possibly making the soil less fertile. Plants are able to excrete H^+ into the soil and by that means, change the pH of the soil near the root and push cations off the colloids, thus making those available to the plant.

Cation Exchange Capacity (CEC)

Cation exchange capacity should be thought of as the soil's ability to remove cations from the soil water solution and sequester those to be exchanged later as the plant roots release hydrogen ions to the solution. CEC is the amount of exchangeable hydrogen cation (H^+) that will combine with 100 grams dry weight of soil and whose measure is one milliequivalents per 100 grams of soil (1 meq/100 g). Hydrogen ions have a single charge and one-thousandth of a gram of hydrogen ions per 100 grams dry soil gives a measure of one milliequivalent of hydrogen ion. Calcium, with an atomic weight 40 times that of hydrogen and with a valence of two, converts to (40/2) x 1 milliequivalent = 20 milliequivalents of hydrogen ion per 100 grams of dry soil or 20 meq/100 g. The modern measure of CEC is expressed as centimoles of positive charge per kilogram (cmol/kg) of oven-dry soil.

Most of the soil's CEC occurs on clay and humus colloids, and the lack of those in hot, humid, wet climates, due to leaching and decomposition respectively, explains the relative sterility of tropical soils. Live plant roots also have some CEC.

Cation exchange capacity for soils; soil textures; soil colloids		
Soil	**State**	**CEC meq/100 g**
Charlotte fine sand	Florida	1.0
Ruston fine sandy loam	Texas	1.9
Glouchester loam	New Jersey	11.9
Grundy silt loam	Illinois	26.3
Gleason clay loam	California	31.6
Susquehanna clay loam	Alabama	34.3
Davie mucky fine sand	Florida	100.8
Sands	------	1 - 5
Fine sandy loams	------	5-10
Loams and silt loams	-----	5-15
Clay loams	-----	15-30
Clays	-----	over 30
Sesquioxides	-----	0-3
Kaolinite	-----	3-15
Illite	-----	25-40
Montmorillonite	-----	60-100
Vermiculite (similar to illite)	-----	80-150
Humus	-----	100-300

Anion Exchange Capacity (AEC)

Anion exchange capacity should be thought of as the soil's ability to remove anions from the soil water solution and sequester those for later exchange as the plant roots release carbonate anions to the soil water solution. Those colloids which have low CEC tend to have some AEC. Amorphous and sesquioxide clays have the highest AEC, followed by the iron oxides. Levels of AEC are much lower than for CEC. Phosphates tend to be held at anion exchange sites.

Iron and aluminum hydroxide clays are able to exchange their hydroxide anions (OH^-) for other anions. The order reflecting the strength of anion adhesion is as follows:

$$H_2PO_4^- \text{ replaces } SO_4^{2-} \text{ replaces } NO_3^- \text{ replaces } Cl^-$$

The amount of exchangeable anions is of a magnitude of tenths to a few milliequivalents per 100 g dry soil. As pH rises, there are relatively more hydroxyls, which will displace anions from the colloids and force them into solution and out of storage; hence AEC decreases with increasing pH (alkalinity).

Soil Reaction (pH)

Soil reactivity is expressed in terms of pH and is a measure of the acidity or alkalinity of the soil. More precisely, it is a measure of hydrogen ion concentration in an aqueous solution and ranges in values from 0 to 14 (acidic to basic) but practically speaking for soils, pH ranges from 3.5 to 9.5, as pH values beyond those extremes are toxic to life forms.

Soil pH

At 25 °C an aqueous solution that has a pH of 3.5 has $10^{-3.5}$ moles H^+ (hydrogen ions) per litre of solution (and also $10^{-10.5}$ mole/litre OH^-). A pH of 7, defined as neutral, has 10^{-7} moles hydrogen ions per litre of solution and also 10^{-7} moles of OH^- per litre; since the two concentrations are equal, they are said to neutralise each other. A pH of 9.5 has $10^{-9.5}$ moles hydrogen ions per litre of solution (and also $10^{-2.5}$ mole per litre OH^-). A pH of 3.5 has one million times more hydrogen ions per litre than a solution with pH of 9.5 ($9.5 - 3.5 = 6$ or 10^6) and is more acidic.

The effect of pH on a soil is to remove from the soil or to make available certain ions. Soils with high acidity tend to have toxic amounts of aluminium and manganese. Plants which need calcium need moderate alkalinity, but most minerals are more soluble in acid soils. Soil organisms are hindered by high acidity, and most agricultural crops do best with mineral soils of pH 6.5 and organic soils of pH 5.5.

In high rainfall areas, soils tend to acidity as the basic cations are forced off the soil colloids by the mass action of hydrogen ions from the rain as those attach to the colloids. High rainfall rates can then wash the nutrients out, leaving the soil sterile. Once the colloids are saturated with H^+, the addition of any more hydrogen ions or aluminum hydroxyl cations drives the pH even lower (more acidic) as the soil has been left with no buffering capacity. In areas of extreme rainfall and high temperatures, the clay and humus may be washed out, further reducing the buffering capacity of the soil. In low rainfall areas, unleached calcium pushes pH to 8.5 and with the addition of exchangeable sodium, soils may reach pH 10. Beyond a pH of 9, plant growth is reduced. High pH results in low micro-nutrient mobility, but water-soluble chelates of those nutrients can correct the deficit. Sodium can be reduced by the addition of gypsum (calcium sulphate) as calcium adheres to clay more tightly than does sodium causing sodium to be pushed into the soil water solution where it can be washed out by an abundance of water.

Base Saturation Percentage

There are acid-forming cations (hydrogen and aluminium) and there are base-forming cations. The fraction of the base-forming cations that occupy positions on the soil colloids is called the base saturation percentage. If a soil has a CEC of 20 meq and 5 meq are aluminium and hydrogen cations (acid-forming), the remainder of positions on the colloids (20-5 = 15 meq) are assumed occupied by base-forming cations, so that the percentage base saturation is 15/20 x 100% = 75% (the compliment 25% is assumed acid-forming cations). When the soil pH is 7 (neutral), base saturation is 100 percent and there are no hydrogen ions stored on the colloids. Base saturation is almost in direct proportion to pH (increases with increasing pH). It is of use in calculating the amount of lime needed to neutralise an acid soil. The amount of lime needed to neutralize a soil must take account of the amount of acid forming ions on the colloids not just those in the soil water solution. The addition of enough lime to neutralize the soil water solution will be insufficient to change the pH, as the acid forming cations stored on the soil colloids will tend to restore the original pH condition as they are pushed off those colloids by the calcium of the added lime.

Buffering of Soils

The resistance of soil to changes in pH, as a result of the addition of acid or basic material, is a measure of the buffering capacity of a soil and increases as the CEC increases. Hence, pure sand has almost no buffering ability, while soils high in colloids have high buffering capacity. Buffering occurs by cation exchange and neutralisation.

The addition of a small amount highly basic aqueous ammonia to a soil will cause the ammonium to displace hydrogen ions from the colloids, and the end product is water and colloidally fixed ammonium, but no permanent change overall in soil pH.

The addition of a small amount of lime, $CaCO_3$, will displace hydrogen ions from the soil colloids, causing the fixation of calcium to colloids and the evolution of CO_2 and water, with no permanent change in soil pH.

The addition of carbonic acid (the solution of CO_2 in water) will displace calcium from colloids, as hydrogen ions are fixed to the colloids, evolving water and slightly alkaline (temporary increase in pH) highly soluble calcium bicarbonate, which will then precipitate as lime ($CaCO_3$) and water at a lower level in the soil profile, with the result of no permanent change in soil pH.

All of the above are examples of the buffering of soil pH. The general principal is that an increase in a particular cation in the soil water solution will cause that cation to be fixed to colloids (buffered) and a decrease in solution of that cation will cause it to be withdrawn from the colloid and moved into solution (buffered). The degree of buffering is limited by the CEC of the soil; the greater the CEC, the greater the buffering capacity of the soil.

Nutrients

Sixteen elements or nutrients are essential for plant growth and reproduction. They are carbon C, hydrogen H, oxygen O, nitrogen N, phosphorus P, potassium K, sulfur S, calcium Ca, magnesium Mg, iron Fe, boron B, manganese Mn, copper Cu, zinc Zn, molybdenum Mo, and chlorine Cl. Nutrients required for plants to complete their life cycle are considered essential nutrients. Nutrients that enhance the growth of plants but are not necessary to complete the plant's life cycle are considered non-essential. With the exception of carbon, hydrogen and oxygen, which are supplied by carbon dioxide and water, the nutrients derive originally from the mineral component of the soil.

Plant uptake of nutrients can only proceed when they are present in a plant-available form. In most situations, nutrients are absorbed in an ionic form from (or together with) soil water. Although minerals are the origin of most nutrients, and the bulk of most nutrient elements in the soil is held in crystalline form within primary and secondary minerals, they weather too slowly to support rapid plant growth. For example, The application of finely ground minerals, feldspar and apatite, to soil seldom provides the necessary amounts of potassium and phosphorus at a rate sufficient for good plant growth, as most of the nutrients remain bound in the crystals of those minerals.

The nutrients adsorbed onto the surfaces of clay colloids and soil organic matter provide a more accessible reservoir of many plant nutrients (e.g. K, Ca, Mg, P, Zn). As plants absorb the nutrients from the soil water, the soluble pool is replenished from the surface-bound pool. The decomposition of soil organic matter by microorganisms is another mechanism whereby the soluble pool of nutrients is replenished - this is important for the supply of plant-available N, S, P, and B from soil.

Gram for gram, the capacity of humus to hold nutrients and water is far greater than that of clay minerals. All in all, small amounts of humus may remarkably increase the soil's capacity to promote plant growth.

Plant nutrients, their chemical symbols, and the ionic forms common in soils and available for plant uptake

Element	Symbol	Ion or molecule
Carbon	C	CO_2 (mostly through leaves)
Hydrogen	H	H^+, HOH (water)
Oxygen	O	O^{2-}, OH^-, CO_3^{2-}, SO_4^{2-}, CO_2
Phosphorus	P	$H_2PO_4^-$, HPO_4^{2-} (phosphates)
Potassium	K	K^+
Nitrogen	N	NH_4^+, NO_3^- (ammonium, nitrate)

Sulfur	S	SO_4^{2-}
Calcium	Ca	Ca^{2+}
Iron	Fe	Fe^{2+}, Fe^{3+} (ferrous, ferric)
Magnesium	Mg	Mg^{2+}
Boron	B	H_3BO_3, $H_2BO_3^-$, $B(OH)_4^-$
Manganese	Mn	Mn^{2+}
Copper	Cu	Cu^{2+}
Zinc	Zn	Zn^{2+}
Molybdenum	Mo	MoO_4^{2-} (molybdate)
Chlorine	Cl	Cl^- (chloride)

Soil Processes Important for Nutrient Uptake

Nutrients in the soil are taken up by the plant through its roots. To be taken up by a plant, a nutrient element must be located near the root surface; however, the supply of nutrients in contact with the root is rapidly depleted. There are three basic mechanisms whereby nutrient ions dissolved in the soil solution are brought into contact with plant roots:

1. Mass flow of water
2. Diffusion within water
3. Interception by root growth

All three mechanisms operate simultaneously, but one mechanism or another may be most important for a particular nutrient. For example, in the case of calcium, which is generally plentiful in the soil solution, mass flow alone can usually bring sufficient amounts to the root surface. However, in the case of phosphorus, diffusion is needed to supplement mass flow. For the most part, nutrient ions must travel some distance in the soil solution to reach the root surface. This movement can take place by mass flow, as when dissolved nutrients are carried along with the soil water flowing toward a root that is actively drawing water from the soil. In this type of movement, the nutrient ions are somewhat analogous to leaves floating down a stream. In addition, nutrient ions continually move by diffusion from areas of greater concentration toward the nutrient-depleted areas of lower concentration around the root surface. That process is due to random motion of molecules. By this means, plants can continue to take up nutrients even at night, when water is only slowly absorbed into the roots as transpiration has almost stopped. Finally, root interception comes into play as roots continually grow into new, undepleted soil.

Estimated relative importance of mass flow, diffusion and root interception as mechanisms in supplying plant nutrients to corn plant roots in soils			
Nutrient	**Approximate percentage supplied by:**		
	Mass flow	**Root interception**	**Diffusion**
Nitrogen	98.8	1.2	0
Phosphorus	6.3	2.8	90.9
Potassium	20.0	2.3	77.7
Calcium	71.4	28.6	0
Sulfur	95.0	5.0	0
Molybdenum	95.2	4.8	0

In the above table, phosphorus and potassium nutrients move more by diffusion than they do by mass flow in the soil water solution, as they are rapidly taken up by the roots creating a concentration of almost zero near the roots (the plants cannot transpire enough water to draw more of those nutrients near the roots). The very steep concentration gradient is of greater influence in the movement of those ions than is the movement of those by mass flow. The movement by mass flow requires the transpiration of water from the plant causing water and solution ions to also move toward the roots. Movement by root interception is slowest as the plants must extend their roots.

Plants move ions out of their roots in an effort to move nutrients in from the soil. Hydrogen H^+ is exchanged for other cations, and carbonate (HCO_3^-) and hydroxide (OH^-) anions are exchanged for nutrient anions. As plant roots remove nutrients from the soil water solution, they are replenished as other ions move off of clay and humus (by ion exchange or desorption), are added from the weathering of soil minerals, and are released by the decomposition of soil organic matter. Plants derive a large proportion of their anion nutrients from decomposing organic matter, which typically holds about 95 percent of the soil nitrogen, 5 to 60 percent of the soil phosphorus and about 80 percent of the soil sulfur. Where crops are produced, the replenishment of nutrients in the soil must usually be augmented by the addition of fertilizer or organic matter.

Because nutrient uptake is an active metabolic process, conditions that inhibit root metabolism may also inhibit nutrient uptake. Examples of such conditions include waterlogging or soil compaction resulting in poor soil aeration, excessively high or low soil temperatures, and above-ground conditions that result in low translocation of sugars to plant roots.

Carbon

Plants obtain their carbon from atmospheric carbon dioxide. About 45% of a plant's dry mass is carbon; plant residues typically have a carbon to nitrogen ratio (C/N) of between 13:1 and 100:1. As the soil organic material is digested by arthropods and micro-organisms, the C/N decreases as the carbonaceous material is metabolized and carbon dioxide (CO_2) is released as a byproduct which then finds its way out of the soil and into the atmosphere. The nitrogen is sequestered in the bodies of the living matter of those decomposing organisms and so it builds up in the soil. Normal CO_2 concentration in the atmosphere is 0.03%, this can be the factor limiting plant growth. In a field of maize on a still day during high light conditions in the growing season, the CO_2 concentration drops very low, but under such conditions the crop could use up to 20 times the normal concentration. The respiration of CO_2 by soil micro-organisms decomposing soil organic matter contributes an important amount of CO_2 to the photosynthesising plants. Within the soil, CO_2 concentration is 10 to 100 times that of atmospheric levels but may rise to toxic levels if the soil porosity is low or if diffusion is impeded by flooding.

Measuring soil respiration in the field using an SRS2000 system.

Nitrogen

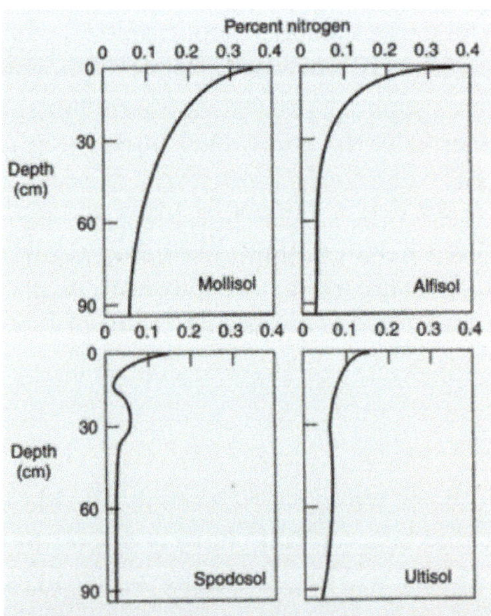

Generalization of percent soil nitrogen by soil order

Nitrogen is the most critical element obtained by plants from the soil and nitrogen deficiency often limits plant growth. Plants can use the nitrogen as either the ammonium cation (NH_4^+) or the anion nitrate (NO_3^-). Usually, most of the nitrogen in soil is bound within organic compounds that make up the soil organic matter, and must be mineralized to the ammonium or nitrate form before it can be taken up by most plants. The total nitrogen content depends largely on the soil organic matter content, which in turn depends on the climate, vegetation, topography, age and soil management. Soil nitrogen typically decreases by 0.2 to 0.3% for every temperature increase by 10 °C. Usually, grassland soils contain more soil nitrogen than forest soils. Cultivation decreases soil nitrogen by exposing soil organic matter to decomposition by microorganisms, and soils under no-tillage maintain more soil nitrogen than tilled soils.

Some micro-organisms are able to metabolise organic matter and release ammonium in a process called *mineralisation*. Others take free ammonium and oxidise it to nitrate. Nitrogen-fixing bacteria are capable of metabolising N_2 into the form of ammonia in a process called nitrogen fixation. Both ammonium and nitrate can be *immobilized* by their incorporation into the microbes' living cells, where it is temporarily sequestered in the form of amino acids and protein. Nitrate may also be lost from the soil when bacteria metabolise it to the gases N_2 and N_2O. The loss of gaseous forms of nitrogen to the atmosphere due to microbial action is called *denitrification*. Nitrogen may also be *leached* from the soil if it is in the form of nitrate or lost to the atmosphere as ammonia due to a chemical reaction of ammonium with alkaline soil by way of a process called *volatilisation*. Ammonium may also be sequestered in clay by *fixation*. A small amount of nitrogen is added to soil by rainfall.

Nitrogen Gains

In the process of mineralisation, microbes feed on organic matter, releasing ammonia (NH_3), ammonium (NH_4^+) and other nutrients. As long as the carbon to nitrogen ratio (C/N) of fresh residues in the soil is above 30:1, nitrogen will be in short supply and other bacteria will feed on the ammonium and incorporate its nitrogen into their cells in the immobilization process. In that form the nitrogen is said to be *immobilised*. Later, when such bacteria die, they too are *mineralised* and some of the nitrogen is released as ammonium and nitrate. If the C/N is less than 15, ammonia is freed to the soil, where it may be used by bacteria which oxidise it to nitrate (nitrification). Bacteria may on average add 25 pounds (11 kg) nitrogen per acre, and in an unfertilised field, this is the most important source of usable nitrogen. In a soil with 5% organic matter perhaps 2 to 5% of that is released to the soil by such decomposition. It occurs fastest in warm, moist, well aerated soil. The mineralisation of 3% of the organic material of a soil that is 4% organic matter overall, would release 120 pounds (54 kg) of nitrogen as ammonium per acre.

Carbon/Nitrogen Ratio of Various Organic Materials	
Organic Material	**C:N Ratio**
Alfalfa	13
Bacteria	4
Clover, green sweet	16
Clover, mature sweet	23
Fungi	9
Forest litter	30
Humus in warm cultivated soils	11
Legume-grass hay	25
Legumes (alfalfa or clover), mature	20
Manure, cow	18
Manure, horse	16–45
Manure, human	10
Oat straw	80
Straw, cornstalks	90
Sawdust	250

In nitrogen fixation, rhizobium bacteria convert N_2 to ammonia (NH_3). Rhizobia share a symbiotic relationship with host plants, since rhizobia supply the host with nitrogen and the host provides rhizobia with nutrients and a safe environment. It is estimated that such symbiotic bacteria in the root nodules of legumes add 45 to 250 pounds of nitrogen per acre per year, which may be sufficient for the crop. Other, free-living nitrogen-fixing bacteria and blue-green algae live independently in the soil and release nitrate when their dead bodies are converted by way of mineralisation.

Some amount of usable nitrogen is fixed by lightning as nitric oxide (NO) and nitrogen dioxide (NO_2^-). Nitrogen dioxide is soluble in water to form nitric acid (HNO_3) solution of H^+ and NO_3^-. Ammonia, NH_3, previously released from the soil or from combustion, may fall with precipitation as nitric acid at a rate of about five pounds nitrogen per acre per year.

Nitrogen Sequestration

When bacteria feed on soluble forms of nitrogen (ammonium and nitrate), they temporarily sequester that nitrogen in their bodies in a process called *immobilisation*. At a later time when those bacteria die, their nitrogen may be released as ammonium by the processes of mineralisation.

Protein material is easily broken down, but the rate of its decomposition is slowed by its attachment to the crystalline structure of clay and when trapped between the clay layers. The layers are small enough that bacteria cannot enter. Some organisms can exude extracellular enzymes that can act on the sequestered proteins. However, those enzymes too may be trapped on the clay crystals.

Ammonium fixation occurs when ammonium pushes potassium ions from between the layers of clay such as illite or montmorillonite. Only a small fraction of soil nitrogen is held this way.

Nitrogen Losses

Usable nitrogen may be lost from soils when it is in the form of nitrate, as it is easily leached. Further losses of nitrogen occur by denitrification, the process whereby soil bacteria convert nitrate (NO_3^-) to nitrogen gas, N_2 or N_2O. This occurs when poor soil aeration limits free oxygen, forcing bacteria to use the oxygen in nitrate for their respiratory process. Denitrification increases when oxidisable organic material is available and when soils are warm and slightly acidic. Denitrification may vary throughout a soil as the aeration varies from place to place. Denitrification may cause the loss of 10 to 20 percent of the available nitrates within a day and when conditions are favourable to that process, losses of up to 60 percent of nitrate applied as fertiliser may occur.

Ammonium volatilisation occurs when ammonium reacts chemically with an alkaline soil, converting NH_4^+ to NH_3. The application of ammonium fertiliser to such a field can result in volatilisation losses of as much as 30 percent.

Phosphorus

After nitrogen, phosphorus is probably the element most likely to be deficient in soils. The soil mineral apatite is the most common mineral source of phosphorus. While there is on average 1000 lb of phosphorus per acre in the soil, it is generally unavailable in the form of phosphates of low solubility. Total phosphorus is about 0.1 percent by weight of the soil, but only one percent of that is available. Of the part available, more than half comes from the mineralisation of organic matter. Agricultural fields may need to be fertilised to make up for the phosphorus that has been removed in the crop.

When phosphorus does form solubilised ions of $H_2PO_4^-$, they rapidly form insoluble phosphates of calcium or hydrous oxides of iron and aluminum. Phosphorus is largely immobile in the soil and is not leached but actually builds up in the surface layer if not cropped. The application of soluble fertilisers to soils may result in zinc deficiencies as zinc phosphates form. Conversely, the application of zinc to soils may immobilise phosphorus again as zinc phosphate. Lack of phosphorus may interfere with the normal opening of the plant leaf stomata, resulting in plant temperatures 10 percent higher than normal. Phosphorus is most available when soil pH is 6.5 in mineral soils and 5.5 in organic soils.

Potassium

The amount of potassium in a soil may be as much as 80,000 lb per acre-foot, of which only 150 lb is available for plant growth. Common mineral sources of potassium are the mica biotite and potassium feldspar, $KAlSi_3O_8$. When solubilised, half will be held as exchangeable cations on clay while the other half is in the soil water solution. Potassium fixation often occurs when soils dry and the potassium is bonded between layers of illite clay. Under certain conditions, dependent on the soil texture, intensity of drying, and initial amount of exchangeable potassium, the fixed percentage may be as much as 90 percent within ten minutes. Potassium may be leached from soils low in clay.

Calcium

Calcium is one percent by weight of soils and is generally available but may be low as it is soluble and can be leached. It is thus low in sandy and heavily leached soil or strongly acidic mineral soil. Calcium is supplied to the plant in the form of exchangeable ions and moderately soluble minerals. Calcium is more available on the soil colloids than is potassium because the common mineral calcite, $CaCO_3$, is more soluble than potassium-bearing minerals.

Magnesium

Magnesium is one of the dominant exchangeable cations in most soils (as are calcium and potassium). Primary minerals that weather to release magnesium include hornblende, biotite and vermiculite. Soil magnesium concentrations are generally sufficient for optimal plant growth, but highly weathered and sandy soils may be magnesium deficient due to leaching by heavy precipitation.

Sulfur

Most sulfur is made available to plants, like phosphorus, by its release from decomposing organic matter. Deficiencies may exist in some soils (especially sandy soils) and if cropped, sulfur needs to be added. The application of large quantities of nitrogen to fields that have marginal amounts of sulfur may cause sulfur deficiency in the rapidly growing plants by the plant's growth outpacing the supply of sulfur. A 15-ton crop of onions uses up to 19 lb of sulfur and 4 tons of alfalfa uses 15 lb per acre. Sulfur abundance varies with depth. In a sample of soils in Ohio, United States, the sulfur abundance varied with depths, 0-6 inches, 6-12 inches, 12-18 inches, 18-24 inches in the amounts: 1056, 830, 686, 528 lb per acre respectively.

Micronutrients

The micronutrients essential for plant life, in their order of importance, include iron, manganese, zinc, copper, boron, chlorine and molybdenum. The term refers to plants' needs, not to their abundance in soil. They are required in very small amounts but are essential to plant health in that most are required parts of some enzyme system which speeds up plants' metabolisms. They are generally available in the mineral component of the soil, but the heavy application of phosphates can cause a deficiency in zinc and iron by the formation of insoluble zinc and iron phosphates. Iron deficiency may also result from excessive amounts of heavy metals or calcium minerals (lime) in the soil. Excess amounts of soluble boron, molybdenum and chloride are toxic.

Non-essential Nutrients

Nutrients which enhance the health but whose deficiency does not stop the life cycle of plants include: cobalt, strontium, vanadium, silicon and nickel. As their importance are evaluated they may be added to the list of essential plant nutrients.

Soil Organic Matter

Soil organic matter is made up of organic compounds and includes plant, animal and microbial material, both living and dead. A typical soil has a biomass composition of 70% microorganisms, 22% macrofauna, and 8% roots. The living component of an acre of soil may include 900 lb of earthworms, 2400 lb of fungi, 1500 lb of bacteria, 133 lb of protozoa and 890 lb of arthropods and algae.

A small part of the organic matter consists of the living cells such as bacteria, molds, and actinomycetes that work to break down the dead organic matter. Were it not for the action of these micro-organisms, the entire carbon dioxide part of the atmosphere would be sequestered as organic matter in the soil.

Chemically, organic matter is classed as follows:

1. Polysaccharides
    - cellulose
    - hemicellulose
    - starch
    - pectin
2. Lignins
3. Proteins

Most living things in soils, including plants, insects, bacteria, and fungi, are dependent on organic matter for nutrients and/or energy. Soils have organic compounds in varying degrees of decomposition which rate is dependent on the temperature, soil moisture, and aeration. Bacteria and fungi feed on the raw organic matter, which are fed upon by amoebas, which in turn are fed upon by nematodes and arthropods. Organic matter holds soils open, allowing the infiltration of air and water, and may hold as much as twice its weight in water. Many soils, including desert and rocky-gravel soils, have little or no organic matter. Soils that are all organic matter, such as peat (histosols), are infertile. In its earliest stage of decomposition, the original organic material is often called raw organic matter. The final stage of decomposition is called humus.

In grassland, much of the organic matter added to the soil is from the deep, fibrous, grass root systems. By contrast, tree leaves falling on the forest floor are the principal source of soil organic matter in the forest. Another difference is the frequent occurrence in the grasslands of fires that destroy large amounts of aboveground material but stimulate even greater contributions from roots. Also, the much greater acidity under any forests inhibits the action of certain soil organisms that otherwise would mix much of the surface litter into the mineral soil. As a result, the soils under grasslands generally develop a thicker A horizon with a deeper distribution of organic matter than in comparable soils under forests, which characteristically store most of their organic matter in the forest floor (O horizon) and thin A horizon.

Humus

Humus refers to organic matter that has been decomposed by soil flora and fauna to the point where it is resistant to further breakdown. Humus usually constitutes only five percent of the soil or less by volume, but it is an essential source of nutrients and adds important textural qualities crucial to soil health and plant growth. Humus also hold bits of undecomposed organic matter which feed arthropods and worms which further improve the soil. The end product, humus, is soluble in water and forms a weak acid that can attack silicate minerals. Humus is a colloid with a high cation and anion exchange capacity that on a dry weight basis is many times greater than that of clay colloids. It also acts as a buffer, like clay, against changes in pH and soil moisture.

Humic acids and fulvic acids, which begin as raw organic matter, are important constituents of humus. After the death of plants and animals, microbes begin to feed on the residues, resulting finally in the formation of humus. With decomposition, there is a reduction of water-soluble constituents, cellulose and hemicellulose, and nutrients such as nitrogen, phosphorus, and sulfur. As the residues break down, only stable molecules made of aromatic carbon rings, oxygen and hydrogen remain in the form of humin, lignin and lignin complexes collectively called humus. While the structure of humus has few nutrients, it is able to attract and hold cation and anion nutrients by weak bonds that can be released into the soil solution in response to changes in soil pH.

Lignin is resistant to breakdown and accumulates within the soil. It also reacts with amino acids, which further increases its resistance to decomposition, including enzymatic decomposition by microbes. Fats and waxes from plant matter have some resistance to decomposition and persist in soils for a while. Clay soils often have higher organic contents that persist longer than soils without clay as the organic molecules adhere to and are stabilised by the clay. Proteins normally decompose readily, but when bound to clay particles, they become more resistant to decomposition. Clay particles also absorb the enzymes exuded by microbes which would normally break down proteins. The addition of organic matter to clay soils can render that organic matter and any added nutrients inaccessible to plants and microbes for many years. High soil tannin (polyphenol) content can cause nitrogen to be sequestered in proteins or cause nitrogen immobilisation.

Humus formation is a process dependent on the amount of plant material added each year and the type of base soil. Both are affected by climate and the type of organisms present. Soils with humus can vary in nitrogen content but typically have 3 to 6 percent nitrogen. Raw organic matter, as a reserve of nitrogen and phosphorus, is a vital component affecting soil fertility. Humus also absorbs water, and expands and shrinks between dry and wet states, increasing soil porosity. Humus is less stable than the soil's mineral constituents, as it is reduced by microbial decomposition, and over time its concentration diminshes without the addition of new organic matter. However, humus may persist over centuries if not millennia.

Climate and Organic Matter

The production, accumulation and degradation of organic matter are greatly dependent on climate. Temperature, soil moisture and topography are the major factors affecting the accumulation of organic matter in soils. Organic matter tends to accumulate under wet or cold conditions where decomposer activity is impeded by low temperature or excess moisture which results in anaerobic conditions. Conversely, excessive rain and high temperatures of tropical climates enables rapid decomposition of organic matter and leaching of plant nutrients; forest ecosystems on these soils rely on efficient recycling of nutrients and plant matter to maintain their productivity. Excessive slope may encourage the erosion of the top layer of soil which holds most of the raw organic material that would otherwise eventually become humus.

Plant Residue in Soil

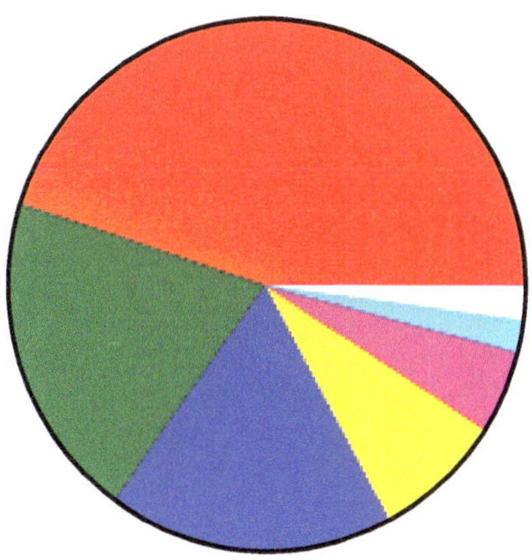

Typical types and percentages of plant residue components

- Cellulose (45%)
- Lignin (20%)

Hemicellulose (18%)
Protein (8%)
Sugars and starches (5%)
Fats and waxes (2%)

Cellulose and hemicellulose undergo fast decomposition by fungi and bacteria, with a half-life of 12–18 days in a temperate climate. Brown rot fungi can decompose the cellulose and hemicellulose, leaving the lignin and phenolic compounds behind. Starch, which is an energy storage system for plants, undergoes fast decomposition by bacteria and fungi. Lignin consists of polymers composed of 500 to 600 units with a highly branched, amorphous structure. Lignin undergoes very slow decomposition, mainly by white rot fungi and actinomycetes; its half-life under temperate conditions is about six months.

Soil Horizons

A horizontal layer of the soil, whose physical features, composition and age are distinct from those above and beneath, are referred to as a soil horizon. The naming of a horizon is based on the type of material of which it is composed. Those materials reflect the duration of specific processes of soil formation. They are labelled using a shorthand notation of letters and numbers which describe the horizon in terms of its colour, size, texture, structure, consistency, root quantity, pH, voids, boundary characteristics and presence of nodules or concretions. Few soil profiles have all the major horizons. Some may have only one horizon.

The exposure of parent material to favourable conditions produces mineral soils that are marginally suitable for plant growth. That growth often results in the accumulation of organic residues. The accumulated organic layer called the O horizon produces a more active soil due to the effect of the organisms that live within it. Organisms colonise and break down organic materials, making available nutrients upon which other plants and animals can live. After sufficient time, humus moves downward and is deposited in a distinctive organic surface layer called the A horizon.

Classification

Soil is classified into categories in order to understand relationships between different soils and to determine the suitability of a soil for a particular use. One of the first classification systems was developed by the Russian scientist Dokuchaev around 1880. It was modified a number of times by American and European researchers, and developed into the system commonly used until the 1960s. It was based on the idea that soils have a particular morphology based on the materials and factors that form them. In the 1960s, a different classification system began to emerge which focused on soil morphology instead of parental materials and soil-forming factors. Since then it has undergone further modifications. The World Reference Base for Soil Resources (WRB) aims to establish an international reference base for soil classification.

Soil Classification Systems

Australia

There are fourteen soil orders at the top level of the Australian Soil Classification. They are: Anthroposols, Organosols, Podosols, Vertosols, Hydrosols, Kurosols, Sodosols, Chromosols, Calcarosols, Ferrosols, Dermosols, Kandosols, Rudosols and Tenosols.

European Union

The EU's soil taxonomy is based on a new standard soil classification in the World Reference Base for Soil Resources produced by the UN's Food and Agriculture Organization. According to this, the major soils in the European Union are:

- Acrisols
- Albeluvisols
- Andosols
- Anthrosols
- Arenosols
- Calcisols
- Cambisols
- Chernozems
- Fluvisols
- Gleysols
- Gypsisols
- Histosols
- Kastanozems
- Leptosols
- Luvisols
- Phaeozems
- Planosols

- Podzols
- Regosols
- Solonchaks
- Solonetzes
- Umbrisols
- Vertisols

USA

A taxonomy is an arrangement in a systematic manner; the USDA soil taxonomy has six levels of classification. They are, from most general to specific: order, suborder, great group, subgroup, family and series. Soil properties that can be measured quantitatively are used in this classification system — they include: depth, moisture, temperature, texture, structure, cation exchange capacity, base saturation, clay mineralogy, organic matter content and salt content. There are 12 soil orders (the top hierarchical level) in the USDA soil taxonomy. The names of the orders end with the suffix *-sol*. The criteria for the different soil orders include properties that reflect major differences in the genesis of soils. The orders are:

- Alfisol - soils with aluminium and iron. They have horizons of clay accumulation, and form where there is enough moisture and warmth for at least three months of plant growth. They constitute 10.1% of soils worldwide.
- Andisol - volcanic ash soils. They are young and very fertile. They cover 1% of the world's ice-free surface.
- Aridisol - dry soils forming under desert conditions which have fewer than 90 consecutive days of moisture during the growing season and are nonleached. They include nearly 12% of soils on Earth. Soil formation is slow, and accumulated organic matter is scarce. They may have subsurface zones of caliche or duripan. Many aridisols have well-developed Bt horizons showing clay movement from past periods of greater moisture.
- Entisol - recently formed soils that lack well-developed horizons. Commonly found on unconsolidated river and beach sediments of sand and clay or volcanic ash, some have an A horizon on top of bedrock. They are 18% of soils worldwide.
- Gelisol - permafrost soils with permafrost within two metres of the surface or gelic materials and permafrost within one metre. They constitute 9.1% of soils worldwide.
- Histosol - organic soils, formerly called bog soils, are 1.2% of soils worldwide.

- Inceptisol - young soils. They have subsurface horizon formation but show little eluviation and illuviation. They constitute 15% of soils worldwide.
- Mollisol - soft, deep, dark fertile soil formed in grasslands and some hardwood forests with very thick A horizons. They are 7% of soils worldwide.
- Oxisol - are heavily weathered, are rich in iron and aluminum oxides (sesquioxides) or kaolin but low in silica. They have only trace nutrients due to heavy tropical rainfall and high temperatures and low CEC of the remaining clays. They are 7.5% of soils worldwide.
- Spodosol - acid soils with organic colloid layer complexed with iron and aluminium leached from a layer above. They are typical soils of coniferous and deciduous forests in cooler climates. They constitute 4% of soils worldwide.
- Ultisol - acid soils in the humid tropics and subtropics, which are depleted in calcium, magnesium and potassium (important plant nutrients). They are highly weathered, but not as weathered as Oxisols. They make up 8% of the soil worldwide.
- Vertisol - inverted soils. They are clay-rich and tend to swell when wet and shrink upon drying, often forming deep cracks into which surface layers can fall. They are difficult to farm or to construct roads and buildings due to their high expansion rate. They constitute 2.4% of soils worldwide.

Alfisol

Andisol

Aridisol

Entisol

Gelisol

Histisol

Inceptisol

Mollisol

Oxisol	Spodosol	Utisol	Vertisol

The percentages listed above are for land area free of ice. "Soils of Mountains", which constitute the balance (11.6%), have a mixture of those listed above, or are classified as "Rugged Mountains" which have no soil.

The above soil orders in sequence of increasing degree of development are Entisols, Inceptisols, Aridisols, Mollisols, Alfisols, Spodosols, Ultisols, and Oxisols. Histosols and Vertisols may appear in any of the above at any time during their development.

The soil suborders within an order are differentiated on the basis of soil properties and horizons which depend on soil moisture and temperature. Forty-seven suborders are recognized in the United States.

The soil great group category is a subdivision of a suborder in which the kind and sequence of soil horizons distinguish one soil from another. About 185 great groups are recognized in the United States. Horizons marked by clay, iron, humus and hard pans and soil features such as the expansion-contraction of clays (that produce self-mixing provided by clay), temperature, and marked quantities of various salts are used as distinguishing features.

The great group categories are divided into three kinds of soil subgroups: typic, intergrade and extragrade. A typic subgroup represents the basic or 'typical' concept of the great group to which the described subgroup belongs. An intergrade subgroup describes the properties that suggest how it grades towards (is similar to) soils of other soil great groups, suborders or orders. These properties are not developed or expressed well enough to cause the soil to be included within the great group towards which they grade, but suggest similarities. Extragrade features are aberrant properties which prevent that soil from being included in another soil classification. About 1,000 soil subgroups are defined in the United States.

A soil family category is a group of soils within a subgroup and describes the physical and chemical properties which affect the response of soil to agricultural management and engineering applications. The principal characteristics used to differentiate soil families include texture, mineralogy, pH, permeability, structure, consistency, the locale's precipitation pattern, and soil temperature. For some soils the criteria also specify the percentage of silt, sand and coarse fragments such as gravel, cobbles and rocks. About 4,500 soil families are recognised in the United States.

A family may contain several soil series which describe the physical location using the name of a prominent physical feature such as a river or town near where the soil sample was taken. An example would be Merrimac for the Merrimack River in New Hampshire, USA. More than 14,000 soil series are recognised in the United States. This permits very specific descriptions of soils.

A soil phase of series, originally called 'soil type' describes the soil surface texture, slope, stoniness, saltiness, erosion, and other conditions.

Uses

Soil is used in agriculture, where it serves as the anchor and primary nutrient base for plants; however, as demonstrated by hydroponics, it is not essential to plant growth if the soil-contained nutrients can be dissolved in a solution. The types of soil and available moisture determine the species of plants that can be cultivated.

Soil material is also a critical component in the mining, construction and landscape development industries. Soil serves as a foundation for most construction projects. The movement of massive volumes of soil can be involved in surface mining, road building and dam construction. Earth sheltering is the architectural practice of using soil for external thermal mass against building walls. Many building materials are soil based.

Soil resources are critical to the environment, as well as to food and fibre production. Soil provides minerals and water to plants. Soil absorbs rainwater and releases it later, thus preventing floods and drought. Soil cleans water as it percolates through it. Soil is the habitat for many organisms: the major part of known and unknown biodiversity is in the soil, in the form of invertebrates (earthworms, woodlice, millipedes, centipedes, snails, slugs, mites, springtails, enchytraeids, nematodes, protists), bacteria, archaea, fungi and algae; and most organisms living above ground have part of them (plants) or spend part of their life cycle (insects) below-ground. Above-ground and below-ground biodiversities are tightly interconnected, making soil protection of paramount importance for any restoration or conservation plan.

The biological component of soil is an extremely important carbon sink since about 57% of the biotic content is carbon. Even on desert crusts, cyanobacteria, lichens and mosses capture and sequester a significant amount of carbon by photosynthesis. Poor farming and grazing methods have degraded soils and released much of this sequestered carbon to the atmosphere. Restoring the world's soils could offset some of the huge increase in greenhouse gases causing global warming, while improving crop yields and reducing water needs.

Waste management often has a soil component. Septic drain fields treat septic tank effluent using aerobic soil processes. Landfills use soil for daily cover. Land application of waste water relies on soil biology to aerobically treat BOD.

Organic soils, especially peat, serve as a significant fuel resource; but wide areas of peat production, such as sphagnum bogs, are now protected because of patrimonial interest.

Geophagy is the practice of eating soil-like substances. Both animals and human cultures occasionally consume soil for medicinal, recreational, or religious purposes. It has been shown that some monkeys consume soil, together with their preferred food (tree foliage and fruits), in order to alleviate tannin toxicity.

Soils filter and purify water and affect its chemistry. Rain water and pooled water from ponds, lakes and rivers percolate through the soil horizons and the upper rock strata, thus becoming groundwater. Pests (viruses) and pollutants, such as persistent organic pollutants (chlorinated pesticides, polychlorinated biphenyls), oils (hydrocarbons), heavy metals (lead, zinc, cadmium), and excess nutrients (nitrates, sulfates, phosphates) are filtered out by the soil. Soil organisms metabolise them or immobilise them in their biomass and necromass, thereby incorporating them into stable humus. The physical integrity of soil is also a prerequisite for avoiding landslides in rugged landscapes.

Degradation

Land degradation refers to a human-induced or natural process which impairs the capacity of land to function. Soils degradation involves the acidification, contamination, desertification, erosion or salination.

Soil acidification is beneficial in the case of alkaline soils, but it degrades land when it lowers crop productivity and increases soil vulnerability to contamination and erosion. Soils are often initially acid because their parent materials were acid and initially low in the basic cations (calcium, magnesium, potassium and sodium). Acidification occurs when these elements are leached from the soil profile by rainfall or by the harvesting of forest or agricultural crops. Soil acidification is accelerated by the use of acid-forming nitrogenous fertilizers and by the effects of acid precipitation.

Soil contamination at low levels is often within a soil's capacity to treat and assimilate waste material. Soil biota can treat waste by transforming it; soil colloids can adsorb the waste material. Many waste treatment processes rely on this treatment capacity. Exceeding treatment capacity can damage soil biota and limit soil function. Derelict soils occur where industrial contamination or other development activity damages the soil to such a degree that the land cannot be used safely or productively. Remediation of derelict soil uses principles of geology, physics, chemistry and biology to degrade, attenuate, isolate or remove soil contaminants to restore soil functions and values. Techniques include leaching, air sparging, chemical amendments, phytoremediation, bioremediation and natural degradation.

Desertification is an environmental process of ecosystem degradation in arid and semi-arid regions, often caused by human activity. It is a common misconception that droughts cause desertification. Droughts are common in arid and semiarid lands. Well-managed lands can recover from drought when the rains return. Soil management tools include maintaining soil nutrient and organic matter levels, reduced tillage and increased cover. These practices help to control erosion and maintain productivity during periods when moisture is available. Continued land abuse during droughts, however, increases land degradation. Increased population and livestock pressure on marginal lands accelerates desertification.

Erosion of soil is caused by water, wind, ice, and movement in response to gravity. More than one kind of erosion can occur simultaneously. Erosion is distinguished from weathering, since erosion also transports eroded soil away from its place of origin (soil in transit may be described as sediment). Erosion is an intrinsic natural process, but in many places it is greatly increased by human activity, especially poor land use practices. These include agricultural activities which leave the soil bare during times of heavy rain or strong winds, overgrazing, deforestation, and improper construction activity. Improved management can limit erosion. Soil conservation techniques which are employed include changes of land use (such as replacing erosion-prone crops with grass or other soil-binding plants), changes to the timing or type of agricultural operations, terrace building, use of erosion-suppressing cover materials (including cover crops and other plants), limiting disturbance during construction, and avoiding construction during erosion-prone periods.

A serious and long-running water erosion problem occurs in China, on the middle reaches of the Yellow River and the upper reaches of the Yangtze River. From the Yellow River, over 1.6 billion tons of sediment flow each year into the ocean. The sediment originates primarily from water erosion (gully erosion) in the Loess Plateau region of northwest China.

Soil piping is a particular form of soil erosion that occurs below the soil surface. It causes levee and dam failure, as well as sink hole formation. Turbulent flow removes soil starting at the mouth of the seep flow and the subsoil erosion advances up-gradient. The term sand boil is used to describe the appearance of the discharging end of an active soil pipe.

Soil salination is the accumulation of free salts to such an extent that it leads to degradation of the agricultural value of soils and vegetation. Consequences include corrosion damage, reduced plant growth, erosion due to loss of plant cover and soil structure, and water quality problems due to sedimentation. Salination occurs due to a combination of natural and human-caused processes. Arid conditions favour salt accumulation. This is especially apparent when soil parent material is saline. Irrigation of arid lands is especially problematic. All irrigation water has some level of salinity. Irrigation, especially when it involves leakage from canals and overirrigation in the field, often raises the underlying water table. Rapid salination occurs when the land surface is within the capillary fringe of saline groundwater. Soil salinity control involves watertable control and flushing with higher levels of applied water in combination with tile drainage or another form of subsurface drainage.

Desertification

Erosion control

Reclamation

Soils which contain high levels of particular clays, such as smectites, are often very fertile. For example, the smectite-rich clays of Thailand's Central Plains are among the most productive in the world.

Many farmers in tropical areas, however, struggle to retain organic matter in the soils they work. In recent years, for example, productivity has declined in the low-clay soils of northern Thailand. Farmers initially responded by adding organic matter from termite mounds, but this was unsustainable in the long-term. Scientists experimented with adding bentonite, one of the smectite family of clays, to the soil. In field trials, conducted by scientists from the International Water Management Institute in cooperation with Khon Kaen University and local farmers, this had the effect of helping retain water and nutrients. Supplementing the farmer's usual practice with a single application of 200 kg bentonite per rai (6.26 rai = 1 hectare) resulted in an average yield increase of 73%. More work showed that applying bentonite to degraded sandy soils reduced the risk of crop failure during drought years.

In 2008, three years after the initial trials, IWMI scientists conducted a survey among 250 farmers in northeast Thailand, half of whom had applied bentonite to their fields. The average improvement for those using the clay addition was 18% higher than for non-clay users. Using the clay had enabled some farmers to switch to growing vegetables, which need more fertile soil. This helped to increase their income. The researchers estimated that 200 farmers in northeast Thailand and 400 in Cambodia had adopted the use of clays, and that a further 20,000 farmers were introduced to the new technique.

If the soil is too high in clay, adding gypsum, washed river sand and organic matter will balance the composition. Adding organic matter (like ramial chipped wood for instance) to soil which is depleted in nutrients and too high in sand will boost its quality.

Pedogenesis

Pedogenesis (from the Greek *pedo-*, or *pedon*, meaning 'soil, earth,' and *genesis*, meaning 'origin, birth') (also termed soil development, soil evolution, soil formation, and soil genesis) is the process of soil formation as regulated by the effects of place, environment, and history. Biogeochemical processes act to both create and destroy order (anisotropy) within soils. These alterations lead to the development of layers, termed soil horizons, distinguished by differences in color, structure, texture, and chemistry. These features occur in patterns of soil type distribution, forming in response to differences in soil forming factors.

Pedogenesis is studied as a branch of pedology, the study of soil in its natural environment. Other branches of pedology are the study of soil morphology, and soil classification. The study of pedogenesis is important to understanding soil distribution patterns in current (soil geography) and past (paleopedology) geologic periods.

Overview

Soil develops through a series of changes. The starting point is weathering of freshly accumulated parent material. Primitive microbes feed on simple compounds (nutrients) released by weathering, and produce acids which contribute to weathering. They also leave behind organic residues. New soils increase in depth by a combination of weathering, and further deposition. Gradually soil is able to support higher forms of plants and animals, starting with pioneer species, and proceeding to more complex plant and animal communities.

Soils deepen with accumulation of organic matter primarily due to the activities of higher plants. Topsoil deepen through soil mixing. Soils develop layers as organic matter accumulates and leaching takes place. This development of layers is the beginning of the soil profile.

A rate of pedogenesis of 1/10 mm per year is in order of magnitude of the global soil production estimates.

Factors of Soil Formation

Russian geologist Vasily Dokuchaev (1846-1903), commonly regarded as the father of pedology, determined in 1883 that soil formation occurs over time under the influence of climate, vegetation, topography, and parent material. He demonstrated this in 1898 using the soil forming equation:

$$\text{soil} = f(cl, o, p)\ \text{tr}$$

(where cl = climate, o = organisms, p = biological processes) tr = relative time (young, mature, old)

Clorpt

In 1941, the American soil scientist Hans Jenny published Factors of Soil Formation. His equation of state differs from the Vasily Dokuchaev equation, treating time (t) as a factor, adding topographic relief (r), and pointedly leaving the ellipsis "open" for more factors (state variables) to be added as our understanding becomes more refined:

$$\text{soil} = f(Cl, o, r, p, t, \ldots)$$

Climate

Climate regulates soil formation. Soils are more developed in areas with higher rainfall and more warmth. The rate of chemical weathering increases by 2-3 times when the temperature increases by 10 degrees Celsius. Climate also affects which organisms are present, affecting the soil chemically and physically (movement of roots).

Organisms

The organisms living in and on the soil form distinct soil types. Coniferous forests have acidic leaf litter that form soils classed as inceptisols. Mixed or deciduous forests leave a larger layer of humus, changing the elements that are either leeched or accumulated in the soil, and thereby forming soils classed as alfisols. Prairies have very high humus accumulation, creating a dark, thick A horizon characteristic of mollisols.

For example, three species of land snails in the genus *Euchondrus* in the Negev desert are noted for eating lichens growing under the surface limestone rocks and slabs (endolithic lichens). They disrupt and eat the limestone. Their grazing results in the weathering of the stones, and the subsequent formation of soil. They have a significant effect on the region: the total population of snails is estimated to process between 0.7 and 1.1 metric ton per hectare per year of limestone in the Negev desert.

Parent Material

The rock from which soil is formed is called parent material. The main types are: aeolian sediments, glacial till, glacial outwash, alluvium, lacustrine sediments and residual parent material (coral or bedrock).

Pedologists see soil formation or soil properties as a function of regional climate, biota, topography, parent material, time and other variables.

Soil Forming Processes

Soils develop from parent material by various weathering processes. Organic matter accumulation, decomposition, and humification are as critically important to soil formation as weathering. The zone of humification and weathering is termed the solum.

Soil acidification resulting from soil respiration supports chemical weathering. Plants contribute to chemical weathering through root exudates.

Soils can be enriched by deposition of sediments on floodplains and alluvial fans, and by wind-borne deposits.

Soil mixing (pedoturbation) is often an important factor in soil formation. Pedoturbation includes churning clays, cryoturbation, and bioturbation. Types of bioturbation include faunal pedoturbation (animal burrowing), floral pedoturbation (root growth, tree-uprootings), and fungal pedoturbation (mycelia growth). Pedoturbation transforms soils through destratification, mixing, and sorting, as well as creating preferential flow paths for soil gas and infiltrating water. The zone of active bioturbation is termed the soil biomantle.

Soil moisture content and water flow through the soil profile support leaching of soluble constituents, and eluviation. Eluviation is the translocation of colloid material, such as organic matter, clay and other mineral compounds. Transported constituents are deposited due to differences in soil moisture and soil chemistry, especially soil pH and redox potential. The interplay of removal and deposition results in contrasting soil horizons.

Key soil forming processes especially important to macro-scale patterns of soil formation are:

- Laterization
- Podsolization
- Calcification
- Salinization
- Gleization

Examples

A variety of mechanisms contribute to soil formation, including siltation, erosion, overpressure and lake bed succession. A specific example of the evolution of soils in prehistoric lake beds is in the Makgadikgadi Pans of the Kalahari Desert, where change in an ancient river course led to millennia of salinity buildup and formation of calcretes and silcretes.

Cover Crop

A cover crop is a crop planted primarily to manage soil erosion, soil fertility, soil quality, water, weeds, pests, diseases, biodiversity and wildlife in an *agroecosystem* (Lu *et al.* 2000), an ecological system managed and largely shaped by humans across a range of intensities to produce food, feed, or fiber. Currently, not many countries are known for using the cover crop method.

Cover crops are of interest in sustainable agriculture as many of them improve the sustainability of agroecosystem attributes and may also indirectly improve qualities of neighboring natural ecosystems. Farmers choose to grow and manage specific cover crop types based on their own needs and goals, influenced by the biological, environmental, social, cultural, and economic factors of the food system in which they operate (Snapp *et al.* 2005). The farming practice of cover crops has been recognized as climate-smart agriculture by the White House.

Soil Erosion

Although cover crops can perform multiple functions in an agroecosystem simultaneously, they are often grown for the sole purpose of preventing soil erosion. Soil erosion is a process that can irreparably reduce the productive capacity of an agroecosystem. Dense cover crop stands physically slow down the velocity of rainfall before it contacts the soil surface, preventing soil splashing and erosive surface runoff (Romkens *et al.* 1990). Additionally, vast cover crop root networks help anchor the soil in place and increase soil porosity, creating suitable habitat networks for soil macrofauna (Tomlin *et al.* 1995). It keeps the enrichment of the soil good for the next few years.

Soil Fertility Management

One of the primary uses of cover crops is to increase soil fertility. These types of cover crops are referred to as “green manure.” They are used to manage a range of soil macronutrients and micronutrients. Of the various nutrients, the impact that cover crops have on nitrogen management has received the most attention from researchers and farmers, because nitrogen is often the most limiting nutrient in crop production.

Often, green manure crops are grown for a specific period, and then plowed under before reaching full maturity in order to improve soil fertility and quality. Also the stalks left block the soil from being eroded.

Green manure crops are commonly leguminous, meaning they are part of the Fabaceae (pea) family.This family is unique in that all of the species in it set pods, such as bean, lentil, lupins and alfalfa. Leguminous cover crops are typically high in nitrogen and can often provide the required quantity of nitrogen for crop production. In conventional farming, this nitrogen is typically applied in chemical fertilizer form. This quality of cover crops is called fertilizer replacement value (Thiessen-Martens *et al.* 2005).

Another quality unique to leguminous cover crops is that they form symbiotic relationships with the rhizobial bacteria that reside in legume root nodules. Lupins is nodulated by the soil microorganism *Bradyrhizobium* sp. (Lupinus). Bradyrhizobia are encountered as microsymbionts in other leguminous crops (*Argyrolobium, Lotus, Ornithopus, Acacia, Lupinus*) of Mediterranean origin. These bacteria convert biologically unavailable atmospheric nitrogen gas (N_2) to biologically available ammonium (NH_{+4}) through the process of biological nitrogen fixation.

Prior to the advent of the Haber-Bosch process, an energy-intensive method developed to carry out industrial nitrogen fixation and create chemical nitrogen fertilizer, most nitrogen introduced to ecosystems arose through biological nitrogen fixation (Galloway *et al.* 1995). Some scientists believe that widespread biological nitrogen fixation, achieved mainly through the use of cover crops, is the only alternative to industrial nitrogen fixation in the effort to maintain or increase future food production levels (Bohlool *et al.* 1992, Peoples and Craswell 1992, Giller and Cadisch 1995). Industrial nitrogen fixation has been criticized as an unsustainable source of nitrogen for food production due to its reliance on fossil fuel energy and the environmental impacts associated with chemical nitrogen fertilizer use in agriculture (Jensen and Hauggaard-Nielsen 2003). Such widespread environmental impacts include nitrogen fertilizer losses into waterways, which can lead to eutrophication (nutrient loading) and ensuing hypoxia (oxygen depletion) of large bodies of water.

An example of this lies in the Mississippi Valley Basin, where years of fertilizer nitrogen loading into the watershed from agricultural production have resulted in a hypoxic "dead zone" off the Gulf of Mexico the size of New Jersey (Rabalais *et al.* 2002). The ecological complexity of marine life in this zone has been diminishing as a consequence (CENR 2000).

As well as bringing nitrogen into agroecosystems through biological nitrogen fixation, types of cover crops known as "catch crops" are used to retain and recycle soil nitrogen already present. The catch crops take up surplus nitrogen remaining from fertilization of the previous crop, preventing it from being lost through leaching (Morgan *et al.* 1942), or gaseous denitrification or volatilization (Thorup-Kristensen *et al.* 2003).

Catch crops are typically fast-growing annual cereal species adapted to scavenge available nitrogen efficiently from the soil (Ditsch and Alley 1991). The nitrogen tied up in catch crop biomass is released back into the soil once the catch crop is incorporated as a green manure or otherwise begins to decompose.

An example of green manure use comes from Nigeria, where the cover crop *Mucuna pruriens* (velvet bean) has been found to increase the availability of phosphorus in soil after a farmer applies rock phosphate (Vanlauwe *et al.* 2000).

Soil Quality Management

Cover crops can also improve soil quality by increasing soil organic matter levels through the input of cover crop biomass over time. Increased soil organic matter enhances soil structure, as well as the water and nutrient holding and buffering capacity of soil (Patrick *et al.* 1957). It can also lead to increased soil carbon sequestration, which has been promoted as a strategy to help offset the rise in atmospheric carbon dioxide levels (Kuo *et al.* 1997, Sainju *et al.* 2002, Lal 2003).

Soil quality is managed to produce optimum circumstances for crops to flourish. The principal factors of soil quality are soil salination, pH, microorganism balance and the prevention of soil contamination.

Water Management

By reducing soil erosion, cover crops often also reduce both the rate and quantity of water that drains off the field, which would normally pose environmental risks to waterways and ecosystems downstream (Dabney *et al.* 2001). Cover crop biomass acts as a physical barrier between rainfall and the soil surface, allowing raindrops to steadily trickle down through the soil profile. Also, as stated above, cover crop root growth results in the formation of soil pores, which in addition to enhancing soil macrofauna habitat provides pathways for water to filter through the soil profile rather than draining off the field as surface flow. With increased water infiltration, the potential for soil water storage and the recharging of aquifers can be improved (Joyce *et al.* 2002).

Just before cover crops are killed (by such practices including mowing, tilling, discing, rolling, or herbicide application) they contain a large amount of moisture. When the cover crop is incorporated into the soil, or left on the soil surface, it often increases soil moisture. In agroecosystems where water for crop production is in short supply, cover crops can be used as a mulch to conserve water by shading and cooling the soil surface. This reduces evaporation of soil moisture. In other situations farmers try to dry the soil out as quickly as possible going into the planting season. Here prolonged soil moisture conservation can be problematic.

While cover crops can help to conserve water, in temperate regions (particularly in years with below average precipitation) they can draw down soil water supply in the spring, particularly if climatic growing conditions are good. In these cases, just before crop planting, farmers often face a tradeoff between the benefits of increased cover crop growth and the drawbacks of reduced soil moisture for cash crop production that season. C/N ratio is balanced with this application.

Weed Management

Thick cover crop stands often compete well with weeds during the cover crop growth period, and can prevent most germinated weed seeds from completing their life cycle and reproducing. If the cover crop is left on the soil surface rather than incorporated into the soil as a green manure after its growth is terminated, it can form a nearly impenetrable mat. This drastically reduces light transmittance to weed seeds, which in many cases reduces weed seed germination rates (Teasdale 1993). Furthermore, even when weed seeds germinate, they often run out of stored energy for growth before building the necessary structural capacity to break through the cover crop mulch layer. This is often termed the *cover crop smother effect* (Kobayashi *et al.* 2003).

Cover crop in South Dakota

Some cover crops suppress weeds both during growth and after death (Blackshaw *et al.* 2001). During growth these cover crops compete vigorously with weeds for available space, light, and nutrients, and after death they smother the next flush of weeds by forming a mulch layer on the soil surface. For example, Blackshaw *et al.* (2001) found that when using *Melilotus officinalis* (yellow sweetclover) as a cover crop in an improved fallow system (where a fallow period is intentionally improved by any number of different management practices, including the planting of cover crops), weed biomass only constituted between 1-12% of total standing biomass at the end of the cover crop growing season. Furthermore, after cover crop termination, the yellow sweetclover residues suppressed weeds to levels 75-97% lower than in fallow (no yellow sweetclover) systems .

In addition to competition-based or physical weed suppression, certain cover crops are known to suppress weeds through allelopathy (Creamer *et al.* 1996, Singh *et al.* 2003). This occurs when certain biochemical cover crop compounds are degraded that happen to be toxic to, or inhibit seed germination of, other plant species. Some well known examples of allelopathic cover crops are *Secale cereale* (rye), *Vicia villosa* (hairy vetch), *Trifolium pratense* (red clover), *Sorghum bicolor* (sorghum-sudangrass), and species in the Brassicaceae family, particularly mustards (Haramoto and Gallandt 2004). In one study, rye cover crop residues were found to have provided between 80% and 95% control of early season broadleaf weeds when used as a mulch during the production of different cash crops such as soybean, tobacco, corn, and sunflower (Nagabhushana *et al.* 2001).

In a recent study released by the Agricultural Research Service (ARS) scientists examined how rye seeding rates and planting patterns affected cover crop production. The results show that planting more pounds per acre of rye increased the cover crop's production as well as decreased the amount of weeds. The same was true when scientists tested seeding rates on legumes and oats; a higher density of seeds planted per acre decreased the amount of weeds and increased the yield of legume and oat production. The planting patterns, which consisted of either traditional rows or grid patterns, did not seem to make a significant impact on the cover crop's production or on the weed production in either cover crop. The ARS scientists concluded that increased seeding rates could be an effective method of weed control.

Disease Management

In the same way that allelopathic properties of cover crops can suppress weeds, they can also break disease cycles and reduce populations of bacterial and fungal diseases (Everts 2002), and parasitic nematodes (Potter *et al.* 1998, Vargas-Ayala *et al.* 2000). Species in the Brassicaceae family, such as mustards, have been widely shown to suppress fungal disease populations through the release of naturally occurring toxic chemicals during the degradation of glucosinolade compounds in their plant cell tissues (Lazzeri and Manici 2001).

Pest Management

Some cover crops are used as so-called "trap crops", to attract pests away from the crop of value and toward what the pest sees as a more favorable habitat (Shelton and Badenes-Perez 2006). Trap crop areas can be established within crops, within farms, or within landscapes. In many cases the trap crop is grown during the same season as the food crop being produced. The limited area occupied by these trap crops can be treated with a pesticide once pests are drawn to the trap in large enough numbers to reduce the pest populations. In some organic systems, farmers drive over the trap crop with a large vacuum-based implement to physically pull the pests off the plants and out of the field (Kuepper and Thomas 2002). This system has been recommended for use to help control the lygus bugs in organic strawberry production (Zalom *et al.* 2001). Another example of trap crops are nematode resistance White mustard (*Sinapis alba*) and Radish (*Raphanus sativus*). They can be grown after a main (cereal) crop and trap nematodes, for example the beet cyst nematode and Columbian root knot nematode. When grown, nematodes hatch and are attracted to the roots. After entering the roots they cannot reproduce in the root due to a hypersensitive resistance reaction of the plant. Hence the nematode population is greatly reduced, by 70-99%, depending on species and cultivation time.

Other cover crops are used to attract natural predators of pests by providing elements of their habitat. This is a form of biological control known as habitat augmentation, but achieved with the use of cover crops (Bugg and Waddington 1994). Findings on the relationship between cover crop presence and predator/pest population dynamics have been mixed, pointing toward the need for detailed information on specific cover crop types and management practices to best complement a given integrated pest management strategy. For example, the predator mite *Euseius tularensis* (Congdon) is known to help control the pest citrus thrips in Central California citrus orchards. Researchers found that the planting of several different leguminous cover crops (such as bell bean, woollypod vetch, New Zealand white clover, and Austrian winter pea) provided sufficient pollen as a feeding source to cause a seasonal increase in *E. tularensis* populations, which with good timing could potentially introduce enough predatory pressure to reduce pest populations of citrus thrips (Grafton-Cardwell *et al.* 1999).

Diversity and Wildlife

Although cover crops are normally used to serve one of the above discussed purposes, they often simultaneously improve farm habitat for wildlife. The use of cover crops adds at least one more dimension of plant diversity to a cash crop rotation. Since the cover crop is typically not a crop of value, its management is usually less intensive, providing a window of "soft" human influence on the farm. This relatively "hands-off" management, combined with the increased on-farm heterogeneity created by the establishment of cover crops, increases the likelihood that a more complex trophic structure will develop to support a higher level of wildlife diversity (Freemark and Kirk 2001).

In one study, researchers compared arthropod and songbird species composition and field use between conventionally and cover cropped cotton fields in the Southern United States. The cover cropped cotton fields were planted to clover, which was left to grow in between cotton rows throughout the early cotton growing season (stripcover cropping). During the migration and breeding season, they found that songbird densities were 7–20 times higher in the cotton fields with integrated clover cover crop than in the conventional cotton fields. Arthropod abundance and biomass was also higher in the clover cover cropped fields throughout much of the songbird breeding season, which was attributed to an increased supply of flower nectar from the clover. The clover cover crop enhanced songbird habitat by providing cover and nesting sites, and an increased food source from higher arthropod populations (Cederbaum *et al.* 2004).

References

- Arthur T. Hubbard, Encyclopedia of Surface and Colloid Science Vol 3, Santa Barbara, California Science Project, Marcel Dekker, New York (2004) ISBN 0-8247-0759-1
- Raynaud, Xavier; Nunan, Naoise (2014). "Spatial ecology of bacteria at the microscale in soil" (PDF). PLOS ONE. 9 (1): e87217. doi:10.1371/journal.pone.0087217
- Chesworth, Ward (2008). Encyclopedia of soil science (PDF). Dordrecht, The Netherlands: Springer. ISBN 978-1-4020-3994-2. Retrieved 18 September 2016
- van der Ploeg, Rienk R.; Schweigert, Peter; Bachmann, Joerg (2001). "Use and misuse of nitrogen in agriculture: the German story" (PDF). Scientific World Journal. 1 (S2): 737–744. doi:10.1100/tsw.2001.263. PMID 12805882
- College of Tropical Agriculture and Human Resources. "Soil Mineralogy". www.ctahr.hawaii.edu. University of Hawai'i. Retrieved 7 August 2014
- Chesworth, Ward (2008). Encyclopedia of soil science (PDF). Dordrecht, The Netherlands: Springer. ISBN 978-1-4020-3994-2. Retrieved 18 September 2016
- Danoff-Burg, James A. "The terrestrial influence: geology and soils". Earth Institute Center for Environmental Sustainability. New York, New York: Columbia University Press. Retrieved 25 September 2016
- Way, J. Thomas (1849). "On the composition and money value of the different varieties of guano". Journal of the Royal Agricultural Society of England. 10: 196–230
- Taylor, Sterling A.; Ashcroft, Gaylen L. (1972). Physical edaphology: the physics of irrigated and nonirrigated soils. San Francisco, California: W.H. Freeman. ISBN 978-0716708186
- Amundson, Ronald. "Soil preservation and the future of pedology" (PDF). Faculty of Natural Resources. Prince of Songkla University, Songkhla, Thailand. Retrieved 9 October 2016
- Way, J. Thomas (1852). "On the power of soils to absorb manure". Journal of the Royal Agricultural Society of England. 13: 123–143
- McCarthy, David F. (2006). Essentials of soil mechanics and foundations: basic geotechnics (7th ed.). Upper Saddle River, New Jersey: Prentice Hall. ISBN 978-0131145603
- "Arizona Master Gardener Manual". Cooperative Extension, College of Agriculture, University of Arizona. Retrieved 4 December 2016
- Hansen, J.; et al. (2008). "Target atmospheric CO2: Where should humanity aim?". Open Atmo-

spheric Science Journal. 2: 217–31. arXiv:0804.1126. Bibcode:2008OASJ....2..217H. doi:10.2174/1874282300802010217

- Gilluly, James; Waters, Aaron Clement; Woodford, Alfred Oswald (1975). Principles of geology (4th ed.). San Francisco, California: W.H. Freeman. ISBN 978-0716702696
- "Soil colloids: properties, nature, types and significance" (PDF). Tamil Nadu Agricultural University. Retrieved 18 December 2016
- Buol, Stanley W.; Southard, Randal J.; Graham, Robert C.; McDaniel, Paul A. (2011). Soil genesis and classification (6th ed.). Ames, Iowa: Wiley-Blackwell. ISBN 978-0470960608
- de Serres, Olivier (1600). Le Théâtre d'Agriculture et mesnage des champs (in French). Paris, France: Jamet Métayer. Retrieved 22 January 2017
- Retallack, Gregory J. (2008). Soils of the past: an introduction to paleopedology (2nd ed.). Hoboken, New Jersey: John Wiley & Sons. ISBN 978-0470698167
- Tandon, Hari L.S. "A short history of fertilisers". Fertiliser Development and Consultation Organisation. Retrieved 5 February 2017
- Miller, Raymond W.; Donahue, Roy Luther (1990). Soils: an introduction to soils and plant growth. Upper Saddle River, New Jersey: Prentice Hall. ISBN 978-0138202262
- Copley, Jon (25 August 2005). "Millions of bacterial species revealed underfoot". New Scientist. Retrieved 19 April 2010
- Fookes, Peter G. (1997). Tropical residual soils. London, United Kingdom: Geological Society of London. ISBN 978-1897799383
- R. B. Brown (September 2003). "Soil Texture" (PDF). Fact Sheet SL-29. University of Florida, Institute of Food and Agricultural Sciences. Retrieved 8 July 2008
- Kononova, M. M. (2013). Soil organic matter: its nature, its role in soil formation and in soil fertility (2nd ed.). Amsterdam, The Netherlands: Elsevier. ISBN 978-1483185682
- "Electrical Design, Cathodic Protection". United States Army Corps of Engineers. 22 April 1985. Archived from the original on 12 June 2008. Retrieved 2 July 2008
- Jelinek, Lawrence J. (1982). Harvest empire: a history of California agriculture. San Francisco, California: Boyd and Fraser. ISBN 978-0878351312

Permissions

All chapters in this book are published with permission under the Creative Commons Attribution Share Alike License or equivalent. Every chapter published in this book has been scrutinized by our experts. Their significance has been extensively debated. The topics covered herein carry significant information for a comprehensive understanding. They may even be implemented as practical applications or may be referred to as a beginning point for further studies.

We would like to thank the editorial team for lending their expertise to make the book truly unique. They have played a crucial role in the development of this book. Without their invaluable contributions this book wouldn't have been possible. They have made vital efforts to compile up to date information on the varied aspects of this subject to make this book a valuable addition to the collection of many professionals and students.

This book was conceptualized with the vision of imparting up-to-date and integrated information in this field. To ensure the same, a matchless editorial board was set up. Every individual on the board went through rigorous rounds of assessment to prove their worth. After which they invested a large part of their time researching and compiling the most relevant data for our readers.

The editorial board has been involved in producing this book since its inception. They have spent rigorous hours researching and exploring the diverse topics which have resulted in the successful publishing of this book. They have passed on their knowledge of decades through this book. To expedite this challenging task, the publisher supported the team at every step. A small team of assistant editors was also appointed to further simplify the editing procedure and attain best results for the readers.

Apart from the editorial board, the designing team has also invested a significant amount of their time in understanding the subject and creating the most relevant covers. They scrutinized every image to scout for the most suitable representation of the subject and create an appropriate cover for the book.

The publishing team has been an ardent support to the editorial, designing and production team. Their endless efforts to recruit the best for this project, has resulted in the accomplishment of this book. They are a veteran in the field of academics and their pool of knowledge is as vast as their experience in printing. Their expertise and guidance has proved useful at every step. Their uncompromising quality standards have made this book an exceptional effort. Their encouragement from time to time has been an inspiration for everyone.

The publisher and the editorial board hope that this book will prove to be a valuable piece of knowledge for students, practitioners and scholars across the globe.

Index